Marc Abraham[illegible] rize Ceremony. He also edits the satirical magazine the *Annals of Improbable Research*, and runs the popular website www.improbable.com. The Ig®Nobel Prizes are awarded every year by genuine Nobel Prize winners to astonished Ig®Nobel Prize winners at a gala ceremony at Harvard University. The Ig®Nobel Prize winners will tour Britain with Marc Abrahams from 12–21 March 2004, during National Science Week.

Marc Abrahams

IG® NOBEL PRIZES

The Annals of Improbable Research

An Orion paperback

First published in Great Britain in 2002
by Orion
This paperback edition published in 2003
by Orion Books Ltd,
Orion House, 5 Upper St Martin's Lane,
London WC2H 9EA

A CIP catalogue record for this book is available from the British Library.

ISBN 0 75284 261 7

Printed and bound in Great Britain by
Clays Ltd, St Ives plc

Contents

Acknowledgements 8

INTRODUCTION 10

What's an Ig? 11

Should You Feel Pleased to Win an Ig Nobel Prize? 14

How the Winners Are Chosen 15

The Ceremony 17

Offensiveness 21

A Message from the V-Chip Monitor 22

A Warning from the V-Chip Monitor 23

Controversy 24

How It Began, Briefly 27

How to Nominate Someone 28

How to Read This Book 31

MEDICAL BREAKTHROUGHS 33

Failure of Electric Shock Treatment for Rattlesnake Envenomation 35

Transmission of Gonorrhea Through an Inflatable Doll 39

Nose Picking in Adolescents 42

Elevator Music Prevents the Common Cold 47

PSYCHOLOGY & INTELLIGENCE 51

A Scholarly Study of Glee 53

A Forbidding Experiment: Spitting, Chewing Gum and Pigeons 57

Ignorance is Bliss 60

ECONOMICS 65

The Economy of Chile versus J.P. Davila 67

Squeezing Orange County/ Bringing Down Barings 70

The Good Lloyd's Shepherds Insure Disaster 74

Dying to Save Taxes 77

PEACE – DIPLOMACY & PERSUASION 81

Parliamentary Punching and Kicking 83

The Levitating Crime Fighters 88

Daryl Gates, the Gandhi of Los Angeles 92

Stalin World 96

PEACEFUL EJACULATIONS & EXPLOSIONS 101

Auto Thieves Flambé 103

Booming Voices of Britain 106

Father of the Bomb 110
Pacific Kaboom 113

LOVE & MARRIAGE 117
The Compulsive Biochemistry of Love 119
Marriage By the Million 124

REPRODUCTION – TECHNIQUES 129
High Velocity Birth 131
Taking Fatherhood in Hand 135

REPRODUCTION – EQUIPMENT 141
Insert Here 143
Height, Penile Length, and Foot Size 150

DISCOVERIES – BASIC SCIENCE 155
A Bug in His Ear 157
The Happiness of Clams 163
To Boldly See What Others Don't 167
Cold Fusion in Chickens 171
Mini-Dinosaurs, Mini-Princesses 175
The Remembrance of Water Passed 183

DISCOVERIES – THINGS THAT FALL OR RISE 189
Injuries Due to Falling Coconuts 191
The Fall of Buttered Toast 195
The Collapse of Toilets in Glasgow 199
Levitating Frogs 203

TROY & THE GRIZZLY BEAR 207
Troy & the Grizzly Bear 209

INVENTIONS 219
The Most Inventive Salesman 221
The Kitty and the Keyboard 226
AutoVision 229
Patenting the Wheel 233

HELLISH TECHNICALITIES 237
Who Is Going to Hell 239
Mikhail Gorbachev is the Antichrist 243

ART & ITS APPRECIATION 251
The Birth of the Plastic Pink Flamingo 253
Pigeons Prefer Picasso? 257
A Good Deed: Trashing Art History 261
Crop Circles 263
Penises of the Animal Kingdom 268

FRAGRANCES – THE GOOD, THE BAD, THE UGLY 275
THE GOOD
The Self-Perfuming Business Suit 277
THE BAD
Filter-Equipped Underwear 281
THE UGLY
A Man Who Pricked His Finger and Smelled Putrid for 5 Years 285

FOOD – PREPARATION & DISPOSAL 289
Extremely Instant Barbecue 291
Bright Blue, and Wiggly 294
Salmonella Excretion in Joy-Riding Pigs 297
Sogginess at Breakfast 301

FOOD – PALATABILITY 305
The Comparative Palatability of Tadpoles 307
The Effects of Ale, Garlic, and Soured Cream on the Appetite of Leeches 313
No Need for Food 319

FOOD – TEA & COFFEE 325
How to Make a Cup of Tea, Officially 327
It Takes Guts: Luak Coffee 332
The Sociology of Canadian Donut Shops 338
The Optimal Way to Dunk a Biscuit 342

EDUCATION 347
Therapeutic Touch 349
Banning the Beaker 354
Deepak Chopra 357
Dan Quayle 362

LITERATURE 365
Farting as a Defence Against Unspeakable Dread 367
The Apostrophe Protection Society 371
976 Co-Authors in Search of a Title 375
948 Titles in Search of an Author 386
Chariots of the Gods 390
The Father of Junk E-Mail 394
Up Theirs 399

APPENDICES 407
Year-by-Year List of Winners 409
The Web Site 428
About the *Annals of Improbable Research* 429

This book is for Robin.

Special Ig thanks to Sid Abrahams, Margot Button, Sip Siperstein, Don Kater, Stanley Eigen, Jackie Baum, Joe Wrinn, Gary Dryfoos, The Harvard Computer Society, The Harvard-Radcliffe Science Fiction Society, and the Harvard-Radcliffe Society of Physics Students.

Special publishing thanks to Regula Noetzli, Trevor Dolby, Pandora White, and Alexa Dalby.

Every year, between 50 and 100 people help organize the Ig Nobel Prize Ceremony. The ceremony and this book would not have been possible without a very lot of help from a very large number of people. Among them: Alan Asadorian and Dorian Photo Lab, Brad Barnhorst, Referee John Barrett, Charles Bergquist, Doug Berman, Silvery Jim Bredt, Blinsky, Alan Brody, Jeff Bryant, Nick Carstoiu, Jon Chase, Keith Clark, Jon Connor, Sylvie Coyaud, Frank Cunningham, Cybercom.net, Investigator T. Divens, Bob Dushman, Kate Eppers, Relena Erskine, Dave Feldman, Len Finegold, Ira Flatow, Stefanie Friedhoff, Jerry Friedman, Martin Gardner, Greg Garrison and the library staff at the Birmingham News, Bruce Gellerman, Sheila Gibson, Shelly Glashow, Margaret Ann Gray, Deborah Henson-Conant, Jeff Hermes, Dudley Herschbach, Holly Hodder, David Holzman, Karen Hopkin, Jo Rita Jordan, Roger Kautz, Hoppin' Harpaul Kohli, Alex Kohn, Deb Kreuze, Leslie Lawrence, Tom Lehrer, Matt Lena, Jerry and Maggie Lettvin, Barbara Lewis, Harry Lipkin, Colonel Bill Lipscomb, Alan Litsky, Julia Lunetta, Counter-Clockwise Mahoney, Lois Malone, Prominent New York attorney William J. Maloney, Mary Chung Restaurant, Micheline Mathews-Roth, Les Frères Michel, The MIT Museum, MIT Press Bookstore, David Molnar, Carol Morton, Lisa Mullins, The Museum of Bad Art, Steve Nadis, Mary O'Grady, Bob Park, Jay Pasachoff, The Flying Petscheks, Stephen Powell, Harriet Provine, Sophie Renaud, Boyce Rensberger, Genevieve Reynolds, Rich Roberts, Nailah Robinson, Nicki Rohloff, Bob Rose, Daniel, Isabelle, Katrina, Natasha, and Sylvia Rosenberg, Louise Sacco, Rob Sanders, the entire magnificent gang at Sanders Theatre, Margo Seltzer, Roland Sharrillo, Sally Shelton, Miles Smith, Smitty Smith, Kris Snibbe, Earle Spamer, Chris Small, Naomi Stephen, Alan Symonds, Judy Taylor, Chris Thorpe, Peaco Todd, Clockwise Twersky, Tom Ulrich, Mark Waldstein, Verena Wieloch, Bob Wilson, Eric Workman, Howard Zaharoff.

And, of course, thank you – on behalf of all of us – to the Ig Nobel Prize winners, and to everyone who has sent in nominations. As we say at the end of each year's ceremony: "If you didn't win an Ig Nobel Prize – and especially if you did – better luck next year!"

Introduction

- What's an Ig?
- Should You Feel Pleased to Win an Ig Nobel Prize?
- How the Winners Are Chosen
- The Ceremony
- Offensiveness
- A Message from the V-Chip Monitor
- A Warning from the V-Chip Monitor
- Controversy
- How It Began, Briefly
- How to Nominate Someone
- How to Read This Book

The Ig Nobel Prize that was awarded in 2000.
Each year the Prize is made from a new, different design.

What's an Ig?

Some people covet it, others flee from it. Some see it as a hallmark of civilization, others as a scuff mark. Some laugh with it, others laugh at it. Many praise it, a few condemn it, others are just mystified. And many people are madly in love with it.

It is the Ig Nobel Prize.

The winners and their achievements are akin to what Sherlock Holmes craved in his famous collection of newspaper clippings:

"He took down the great book in which, day by day, he filed the agony columns of the various London journals. 'Dear me!' said he, turning over the pages, 'what a chorus of groans, cries, and bleatings! What a ragbag of singular happenings! But surely the most valuable hunting ground that ever was given to a student of the unusual!'"

Sherlock Holmes was, of course, fictional. The Ig Nobel Prize winners are not.

Each year ten Ig Nobel Prizes are awarded to people whose achievements "cannot or should not be reproduced". The

"Igs" (as they are known) honor people who have done remarkably goofy things – some admirable, some perhaps otherwise.

Each of the Prize-winning achievements makes people: (a) laugh and (b) shake their heads in wonder.

Many are scientific. Norwegian biologists who evaluated the effects of ale, garlic, and soured cream on the appetite of leeches. An American professor who fed Prozac to clams. A New York veterinarian who inserted ear mites from a cat into his own ear, and carefully noted what happened after that. A French biologist whose experiments show that water can remember things. A Canadian professor who arranged a taste test of – not for – different species of Costa Rican tadpoles.

Some are gustatory. A British physicist who determined the optimal way to dunk a biscuit. An Australian motivational speaker who explains that people don't really need to eat food. A Korean religious leader who brought steady growth to the million-couples-at-a-time marriage industry.

Some are economic. A man who caused the collapse of Britain's most venerable bank. A Chilean commodities trader who lost 0.5% of his country's gross national product. American economists who showed that death and taxes have an oddly intimate relationship. The father of the junk bond. The curiously tangled investors of Lloyd's of London.

Some concern discovery and loss. An amateur scientist who discovered ten-mile-high buildings on the back side of the moon. French boy scouts who wiped away ancient cave paintings, believing them to be a form of graffiti. Surgeons who compiled a comprehensive historical survey of objects found in people's rectums.

Some are medical. A man who pricked his finger and

smelled putrid for five years, and the doctors who treated him. Canadian doctors who studied the relationship between height, penile length, and foot size. Indian psychiatrists who discovered that nose picking is a common activity among adolescents. Scottish physicians who documented the collapse of toilets in Glasgow.

Some involve the propagation of the species. A Dutch research team that took the first MRI pictures of a couple's genitals while those genitals were in use. A childless, elderly couple who invented a machine that uses centrifugal force to help women give birth.

Some involve art. The creator of the plastic pink flamingo. The man behind the classic anatomy poster *Penises of the Animal Kingdom*. The psychology professors who discovered that listening to elevator music helps prevent illness. Japanese psychologists who trained pigeons to discriminate between the paintings of Picasso and those of Monet.

Some are literary. The 976 co-authors of a ten-page scientific paper. An Italian psychologist who authored the report "Farting as a Defence Against Unspeakable Dread." A Philadelphia businessman who made electronic junk mail ubiquitous. A retired editor who founded The Apostrophe Protection Society.

And some achievements earn the coveted Ig Nobel Peace Prize. National leaders who set off atomic bombs in each other's back yards. The British Navy, which told its sailors to stop using artillery shells and instead just say "Bang." A Lithuanian mushroom tycoon who created the amusement park known as Stalin World. A South African husband-and-wife team who invented an automobile burglar alarm that consists of a motion detector and a flame-thrower.

And many more.

These things can be difficult to believe. That is why the Ig Nobel Board of Governors publishes information that anyone can use to verify and savor the details.

That is also why the winners are invited to come to the Ig Nobel Prize Ceremony, which is held each October at Harvard University. The winners must travel at their own expense, and for many it is apparently worth the cost. A friendly standing-room-only audience of 1,200 welcomes them with warm, wild applause and paper airplanes.

In a unique ritual, genuine Nobel Laureates physically hand the Ig Nobel Prizes to the new Ig Nobel Prize winners. Each time this occurs it is a magical instant – at that moment it feels as if the universe has two opposite ends, and these two opposite ends have somehow managed to meet and touch. Nobel Laureate and Ig Nobel Laureate look each other in the eye, each filled with gleeful wonder.

Should You Feel Pleased to Win an Ig Nobel Prize?

Lucky, yes. Certainly you should feel lucky. Some people spend years striving to win an Ig – yet they never get one. For most winners, the honor is something that sneaked or caught up on them. One fine day came the notification, followed up by several conversations to make sure that, yes, it has arrived at last – official recognition by a bemused and curious world that they have done something, well, curious and bemusing.

Most of the Ig Nobel Prize winners have, in fact, been pleased, at least a little bit, at the news. True, it is a curious honor, but life is short and, well, why not accept it?

Most other kinds of prizes extol the good, or mock the bad. The world, in general, seems to enjoy classifying things as being either one or the other. Olympic medals go to very good athletes. "Worst Dressed" prizes go to badly dressed celebrities. Nobel Prizes go to scientists, writers, and others who excel. Occasional mistakes and omissions happen, sure, but those prizes, and most others, are meant to honor the extremes of humanity – those whose achievements should be seen as very good or very bad.

The Ig Nobel Prize isn't like that. The Ig (as it is known) honors the great muddle in which most of us exist much of the time. Life is confusing. Good and bad get all mixed up. Yin can be hard to distinguish from yang. Ditto for forest from tree and, sometimes, up from down.

Most people go through life without ever being awarded a great, puffy prize to acknowledge that, yes, they have done something. That's why we award Ig Nobel Prizes. If you win one, it signifies to one and all that you have done something. What that thing is may be hard to explain – may even be totally inexplicable. Whether your achievement is for the public good or bad may be difficult, or even painful, to explain. But the fact is, you did it and have been recognized for doing it. Let others make of that recognition what they will.

How the Winners Are Chosen

It cannot be overstated that the Ig Nobel Prize winners and their accomplishments are real.

At the conclusion of one Ig Nobel Prize Ceremony, a female journalist from England climbed onto the stage and accosted a Nobel Laureate who had just helped hand out the awards.

"This was your first Ig, wasn't it?" she asked the distinguished scientist. "Did you enjoy it?" "Oh, yes," he said, eyes crinkling in delight. "Those people were so funny! Can you imagine if they'd really done those things?"

The reporter gave a low chuckle. "They *did* do those things."

Who chooses the winners? The Ig Nobel Board of Governors. Who are the Ig Nobel Board of Governors? Ah. The group comprises the editors of the *Annals of Improbable Research* (the science humor magazine I edit), and a considerable number of scientists (including, yes, several Nobel Laureates), journalists, and others in a variety of fields in a variety of countries. The group never meets at a single gathering. We keep no records of who sent in nominations or, for that matter, of who exactly is on the committee. There is a tradition that for the final decision, we grab some passer-by from the street, to add a little balance.

Where do the nominations come from? Anywhere. Everywhere. Anyone can nominate anyone for an Ig Nobel Prize, and pretty much anyone does. We receive several thousand nominations each year, among them quite a few persons nominating themselves (to date, only one Prize has ever been awarded to a self-nominee – the Norwegian team of Baerheim and Sandvik, who researched the effects of ale, garlic, and soured cream on the appetite of leeches).

Generally, those who are selected can turn down the Prize if they truly believe it might cause them professional difficulties with bosses, governments, or the like. But in the more-than-ten years that Prizes have been given, only a small handful have declined. In recent years most winners have chosen to attend the ceremony or, when finances or other circumstances made that difficult, at least send an acceptance speech.

The winners who come to collect their Prizes always receive a warm welcome. If someone is sporting enough to come celebrate their goofy achievement in public, the audience and the organizers always give them an appreciative, if chuckling, tip of the cap.

The Ceremony

If you win an Ig Nobel Prize, the best part is that you get to star in the Ig Nobel Prize Ceremony, and be the upside-down-buttered toast of the town.

The ceremony began as something giddy in the dead of night, with 350 people crammed inside a museum at the Massachusetts Institute of Technology. That first year, 1991, we invited four Nobel Laureates to come help hand out the Prizes. All four showed up, wearing Groucho glasses, sashes, fezzes, and other stylishly sportive attire. The public was invited to attend, and almost instantly snapped up all the tickets. Reporters came, too, and on that evening everyone had the giddy feeling of sneaking something really different into being. The emphasis here is on the word "sneak," because we all felt as though sooner or later some authority figure would rush in and tell us to stop this nonsense and go home. But no one did, it was a wild success, and the next year we had to move it to the largest meeting place at MIT.

Thereafter, nominations came in a never-ending flood, and every year, spectators, winners, and Nobel Laureates came from great distances to take part in the ceremony.

After the Fourth First Annual Ig Nobel Prize Ceremony in 1994, a dyspeptic MIT administrator tried to ban the event. Puzzled, but almost amused, the Ig Nobel Board of Governors simply moved everything two miles up the road, where it now has a permanent home at Sanders Theatre, Harvard University's oldest, largest and most stately meeting place.

Several Harvard student groups co-sponsor the event together with the *Annals of Improbable Research*. Many Harvard and MIT faculty members, students, and administrators, and many other people as well, are part of what is now a year-round, all-volunteer organizing effort.

The ceremony itself has grown ever more complex, a jaunty meld of every dignified convention with its every off-balance antidote, heaped high with essence of Academy Awards, coronation, circus, football game, opera, booby hatch, laboratory accident, and the old Broadway show *Hellzapoppin'*. Each year more goodies are jammed around, between, and atop the awarding of the ten new Ig Nobel Prizes. My role as master of ceremonies has been likened to that of Kermit the Frog, trying desperately to keep some thread of calm and dignity in a theater filled with brilliant lunatics, each swinging full tilt through his or her own independent universe.

A tradition sprang up in about the second year, whereby the audience members – all 1,200 of them – spend the entire evening wafting paper airplanes at the stage, and the people on stage spend the evening wafting them right back. The volume of paper dropping onto the stage is so great that we detail two people to constantly sweep away detritus; without this it would be nigh impossible to get about the stage.

The evening begins with the traditional "Welcome, Welcome" speech, delivered by an elderly matriarch and comprising in its entirety the statement, "Welcome, welcome." There is a grand and motley entrance parade of audience delegations such as the Museum of Bad Art; the Lawyers for and Against Complexity; the Society for the Preservation of Slide Rules; the Junior Scientists' Club (all of whose members are about seven years old); Fruitcakes for a Better Tomorrow; the Society of Bearded Men; the

Harvard Bureaucracy Club; Grannies Against Gravity; and the protest group Non-Extremists for Moderate Change in Finland.

At some point in the evening comes the Win-a-Date-With-a-Nobel-Laureate Contest, in which one lucky audience member wins a date with a Nobel Laureate.

The 1994 ceremony included the world premiere and only performance of *The Interpretive Dance of the Electrons*, a ballet performed by the Nicola Hawkins Dance Company and starring Nobel Laureates Richard Roberts, Dudley Herschbach, and William Lipscomb.

Every year since 1996 we have written a mini-opera which is then performed by professional opera singers and several Nobel Laureates. The key to making these operas work is to cast them with a mix of performers, all of whom are either (a) extremely skilled and talented or (b) endearingly game. *The Cockroach Opera* was our first. Later years saw the premieres of *Il Kaboom Grosso* (about the Big Bang, with a denouement featuring five Nobel Laureates as subatomic particles), *The Seedy Opera* (starring five tenors playing the role of Ig Nobel Prize-winning physicist Richard Seed, the man who plans to clone himself), and other musical delights.

Each year's ceremony also includes some special event in which celebrities from the worlds of science, literature, and art get to show off unexpected talents.

The Heisenberg Certainty Lectures (named after the famous Heisenberg Uncertainty Principle, which was named after Nobel Laureate Werner Heisenberg) gave many renowned scientists, university presidents, actors, politicians, and musicians the opportunity to lecture the audience on any topic they wished, with no restrictions save one. Each Heisenberg Lecturer was strictly limited to thirty seconds, with the time limit enforced by a professional soccer referee. Anyone who exceeded the time limit was thrown off the stage. This proved popular with the audience.

One year, a collection of celebrated thinkers engaged in a contest to determine which of them is the world's smartest person. This was decided in a series of one-on-one, 30-second-long debates in which both debaters had to talk at the same time. Here, too, our referee Mr John Barrett enforced the time limit.

At both the Sixth and Seventh First Annual Ig Nobel Prize Ceremonies, we auctioned off plaster casts of the (left) feet of Nobel Laureates. The proceeds were donated to the science programs of local schools.

The Eleventh First Annual Ceremony culminated in a wedding – a genuine wedding – of two scientists. The wedding ceremony was 60 seconds long, with 1,200 guests, including four teary-eyed Nobel Laureates and 40 people wearing Joseph Stalin masks (it's a long story, related to that year's Peace Prize winner), the whole thing televised live on the Internet. Ig Nobel Prize winner Buck Weimer, the inventor of airtight underwear with a replaceable charcoal filter that removes bad-smelling gases before they can escape, presented the newlyweds with pairs of the underwear and instructed them on its use. Late that night, as the bride's mother was leaving Sanders Theatre, she beamingly told everyone that "This wasn't exactly what I would have planned for my daughter ... but it was even better."

Every year, with so much going on during the Ceremony, and with so many people having to give speeches, we faced a severe problem: how to graciously stop anyone who couldn't, or wouldn't, keep it brief. Our success with the 30-second-long Heisenberg Certainty Lectures eventually led us to an overall solution, and in 1999 we introduced a great technical innovation called "Miss Sweetie Poo."

Miss Sweetie Poo is an exceptionally cute eight-year-old girl. Whenever Miss Sweetie Poo feels that a speaker has exceeded his or her allotted time, she walks up to the lectern, looks up at the speaker, and says, "Please stop. I'm bored. Please stop. I'm bored. Please stop. I'm bored." Miss Sweetie Poo keeps saying this until the speaker gives up.

Miss Sweetie Poo is *very* effective. Since she has been part of the show, the ceremony has been 40% briefer than it had been before. Miss Sweetie Poo is our greatest invention.

Press coverage of the Ig from virtually every country on earth has grown and grown, and we have tried to make it easy for people in distant places to get a glimpse of the ceremony. Every year since 1993, National Public Radio broadcasts the Ig across North America, and ever since the Fifth 1st Annual Ig, in 1995, we have telecast every ceremony live on the Internet. For several years, our telecast engineer was Harvard graduate student and convicted felon Robert Tappan Morris, whose worm program brought down the entire Internet and made him the first celebrated cyberspace criminal. You can see video and other highlights at the *Annals of Improbable Research* web site (www.improbable.-com).

And you are invited to send in a nomination, should you happen to know someone deserving, for one of next year's Ig Nobel Prizes.

Offensiveness

At the Ig Nobel Ceremony, we work hard to maintain a tone of dignity and propriety, twisted three degrees counter-clockwise.

The audience has definite sensibilities, though we are never sure what they are. There are young children and elderly great-grandparents, and clergy-persons, and sensitive scientists watching and listening – some seated in the theater, some via the Internet telecast, and others glued or otherwise fastened to the radio.

To ensure that nothing offensive reaches the eyes, ears, or (on the Internet) the fingertips of the audience, we employ someone to Keep Things in Line.

Mr William J. Maloney is a prominent New York attorney. Every year he takes a day off from work, and drives the 250 miles north to Cambridge, Massachusetts, where he fulfills his duties as V-Chip Monitor. Armed with a cheap tin bugle, a little flag, a well-cut business suit, and enormous personal dignity, the V-Chip Monitor puts a halt to anything he deems potentially offensive.

The Ig Nobel Board of Governors asked Mr Maloney to perform his duties for readers of this book. Here is a message from Mr Maloney.

A Message from the V-Chip Monitor

"As the V-Chip Monitor, or censor, if you prefer, of the Ig Nobel Ceremony, it is my duty, and singular pleasure, to prevent the audience from seeing and hearing that which they most want to see and hear. My occasional interventions are typically greeted with hostility by a vocal minority who, like the baby from whom one has taken candy, believe I have deprived them of some particularly wonderful entertainment.

"But would they have been entertained, or repulsed, if I had permitted the demonstration of the transmission of gonorrhea through an inflatable doll which I prophylactically prevented at the 1996 Ceremony? Did the audience at the 2001 Ceremony need to smell for themselves whether Buck Weimer's Under-Ease airtight underpants really do remove bad-smelling gases before they escape? Are we amused by the sight of distinguished scientists, Nobel Laureates, picking their noses with gigantic artificial appendages, or demonstrating the relationship between height, penile length, and foot size, again with artificial appendages of grotesque proportions? Does a civilized society derive pleasure from observing living subjects' pain experienced during execution by different methods, a demonstration planned to commemorate the 1997 Peace Prize? Indeed, do we really need to observe a fat man collapse a toilet to know that it can be done?

"This is but a partial list of the potential perversities from which I have shielded the audience. Upon reflection, I am sure that even my most ardent critics would agree that my vigilance is absolutely necessary to preserve the dignity of the Ig Nobel Awards Ceremony. While Science is pure and wholesome, those who pursue it, especially those of Ig Nobel achievement, seem increasingly predisposed to impure thought and action. Fortunately, the V-Chip Monitor stands ready to block anything offensive from reaching your eyes, ears or fingertips."

A Warning from the V-Chip Monitor

"A cursory examination of this book leads me to conclude that it is not suitable for readers of any age. It is replete with offensive, repugnant images, words, phrases and ideas. I urge you not to purchase it. If you have purchased it, do not read it. Should it become known that you possess this book, you will be subject to ridicule and social ostracization.

With considerable effort the book may be rendered innocuous, though it will still not be a particularly good book. You should remove the following pages in their entirety ■■■■■. The images which appear on pages ■■■■■■■■■■ should be covered with tape. For my personal library I enjoy using duct tape for this purpose – it is opaque and any attempt to remove the tape will destroy the underlying image."

Controversy

The Igs have not been without controversy. Sir Robert May, the science adviser to the British government, asked the organizers to stop giving Ig Nobel Prizes to British scientists – even when the scientists want to receive them. May sent two angry letters to the Ig Nobel Board of Governors, and later granted interviews to the press. The reaction was not what he was expecting. Typical was the following editorial, which appeared in the October 7, 1996 issue of the British science journal *Chemistry & Industry*. It is reprinted here with permission from *Chemistry & Industry*.

WE ARE AMUSED

Is Britain's chief scientific adviser, Robert May, a pompous killjoy? In his recently publicised criticism of the Ig Nobel awards, a well-established spoof of the Nobel Prizes, he appears only to confirm that the British scientific establishment takes itself far too seriously.

In an interview with the journal *Nature*, May warns that the Ig Nobels risk bringing "genuine" scientific projects into counter-productive ridicule. They should focus on anti-science and pseudo-science, he suggests, "while leaving serious scientists to get on with their work." His pique stems from embarrassing media coverage given to UK food scientists after an award last year for their research on soggy cereal flakes.

Such whining has several flaws. First, it is not for bureaucrats like May to determine which scientists are "serious," or to ask that some researchers be ignored because they are above being made fun of (they aren't – the good ones as well as the bad ones).

Secondly, the Ig Nobels are organised by academics, for academics – unlike the notorious Golden Fleece awards in the US, with which May compares the Ig Nobels. The Ig Nobels let science laugh at itself.

Thirdly, the work of genuinely "serious" scientists will withstand transitory embarrassment at the hands of TV comics and tabloid newspapers – assuming, of course, that their work really is recognised as "serious" by other scientists. If, under a sudden spotlight, some scientists have to spend much time and effort explaining to everyone why their work is worth funding, that is a good thing and should happen more often, not less.

Finally, May reportedly suggests that the Ig Nobel organiser should obtain winners' consent first. But the British scientists did agree to receive their award last year, which makes May's grumbling distinctly off-target. Furthermore, that particular award proved that media mischief can not be avoided by obtaining prior consent. As the Ig Nobel organiser, Marc Abrahams, has pointed out to May, "there are few things, good or bad, that British tabloids and TV comedians do not ridicule."

> Far from making a convincing case for the pernicious effect of the Ig Nobels, May's misfire only makes him (and British science) look thin-skinned and humourless. He mistakes discomfort for disaster, and solemnity for seriousness. And he misunderstands the point, the process, and the pleasure of the awards. On this topic, scientists and others should reject this adviser's ill-advised views. Long may British scientists take their rightful places in the Ig Nobel honour roll.

The 1995 Prize that prompted May's complaint honored three Norwich scientists "for their rigorous analysis of soggy breakfast cereal, published in a report titled 'A Study of the Effects of Water Content on the Compaction Behaviour of Breakfast Cereal Flakes.'" That same year, Nick Leeson won a share of the Economics Prize for his role in bringing down Barings Bank.

The Robert May flap did not deter the Ig Nobel Board of Governors from giving full consideration to high achievers in the United Kingdom. Nor did it deter future winners from accepting their unusual place on the world stage.

In 1996, undaunted by the very public stance of his nation's chief science official, Robert Matthews of Aston University won, and happily accepted, the Physics Prize for demonstrating that toast often falls on the buttered side. In 1998, three doctors from the Royal Gwent Hospital shared the Medicine Prize with the anonymous patient who was the subject of their cautionary medical report, "A Man Who Pricked His Finger and Smelled Putrid for 5 Years." Indeed, the UK has produced at least one winner (and often more) every year since 1992. The pool of UK nominees for the Ig Nobel Prize is so deep that it could easily supply all ten winners every year. But the same is true of many other nations. In the unending competition for Igs, reputation alone counts for nothing. No country can, or should, rely on the glories of its past accomplishments.

How It Began, Briefly

The Ig Nobel Prize Ceremony was born not long after I unexpectedly became the editor of a magazine called the *Journal of Irreproducible Results*. The *Journal* was started in 1955 by Alex Kohn and Harry Lipkin, two eminent and very funny scientists in Israel, but it eventually fell into other hands and withered to near-extinction. In 1990, I mailed off some articles to see whether this journal (which I had never seen) still existed, and if so whether it might print them. Several weeks later came a telephone message from a man who said he was the publisher, that he'd gotten the articles, and would I be the magazine's editor.

As the editor of a science magazine, even a funny one, I was besieged by people who wanted my help in winning a Nobel Prize. I always explained that I had no influence on these matters, but they invariably told me in great detail what they'd done and why they deserved a prize. In some cases, they were right. They deserved a prize, but not a Nobel Prize.

And so, together with everyone I could talk into helping out, I started the Ig Nobel Prize Ceremony. Alex Kohn suggested naming it after the "Ignoble Prize," a fictional award he and Harry Lipkin had described years before.

The Ig Nobel Prizes would be awarded to people who had done something spectacularly goofy and thought-provoking – "achievements that cannot or should not be reproduced." Some of these achievements might be wonderfully goofy, others appallingly so. Some might even turn out to be – who knows? – both goofily good and maybe even important. Some prizes would be for scientific achievements, others for economics, peace, or other fields. We chose seven winners and invited them to come to the ceremony (but we were new at this, and were only able to reach a few of them). We also, that first year, chose three apocryphal achievements.

We held the first Ig Nobel Prize Ceremony in October 1991. Of the ten winners, it immediately became clear that the genuine achievers were by any measure better than the concocted ones. And so, in all subsequent years, the Ig Nobel Prizes have been awarded always to real people for genuine, documented accomplishments.

The publishers of that magazine, by the way, had a corporate shuffle, and made it clear that a science humor magazine was no longer for them. Rather than watch it go down the tubes, we all left and immediately started a new magazine, the *Annals of Improbable Research* (*AIR*). I have fond memories of the day when four Nobel Laureates each separately informed me, with appropriate cackles, that I am an *AIR*head. *AIR* is the proud home base of the Ig.

How to Nominate Someone

Official Criterion for winning a Prize

Ig Nobel Prizes are given for "achievements that cannot or should not be reproduced."

Unofficial Criterion for winning a prize

A winning achievement must be both goofy and thought-provoking.

Who is authorized to send in Nominations
Anyone.

Who is eligible to Win
Anyone, anywhere. All sorts of people get unusual ideas, and vow to act on them. Those destined to win an Ig Nobel Prize get very unusual ideas, and don't bother to make vows – they simply swing into action. Shoes and ships; cabbages and kings; centrifugal birthing machines and leech appetite stimulants; the drafting of comprehensive technical specifications for making a cup of tea; the classification of foreign objects found in the rectums of medical patients – any of these could be the basis for an Ig Nobel Prize-winning achievement. And most of them have been. You can nominate a stranger, a colleague, a boss, a spouse, or yourself. You can nominate an individual or a group.

Who is not eligible to win
People who are fictional or whose existence – and achievement – cannot be verified.

Categories
Once the winners are selected, each Prize is given in a particular category. Some categories recur every year – Biology, Medicine, Physics, Peace, Economics. Other categories (Safety Engineering, Environmental Protection) are created to fit the particular and/or peculiar nature of a particular achievement. But, in truth, it is impossible to confine Ig Nobel Prize winners within categories. (It is not, however, impossible to confine Ig Nobel Prize winners. Many winners of the Economics Prize, for example, have been unable to attend the Ig Nobel Prize Ceremony because they were serving a prior engagement of five to fifteen years.)

Goodness or badness

Every year, of the ten new Ig Nobel Prizes, about half are awarded for things that most people would say are commendable – if perhaps goofy. The other half go for things that are, in some people's eyes, less commendable. All judgments as to "goodness" and/or "badness" are entirely up to each observer.

How to send Information

Gather information that explains *who* the nominee is, and *what* the nominee has accomplished. Please include enough information that the judges can get an immediate, clear appreciation of why the candidate deserves an Ig Nobel Prize. Also indicate where the judges can find further information if they need it, including (if you know it) how we can get in touch with the nominee. Mail or e-mail the nomination to:

IG NOBEL NOMINATIONS
C/O ANNALS OF IMPROBABLE RESEARCH
PO BOX 380853
CAMBRIDGE MA 02238 USA
E-mail: air@improbable.com

If you mail the material and would like a response, please include an e-mail address or an adequately stamped, self-addressed envelope. If you wish anonymity, you can have it. The Ig Nobel Board of Governors typically loses or discards most of its records, in any event.

You can find further information at the *Annals of Improbable Research* web site (www.improbable.com).

How to Read This Book

This book is meant to be read aloud, preferably in elevators for the edification of your fellow passengers. Trains, buses, subways, and waiting rooms are other good places. If you work in a group that has a tedious weekly meeting, try reading one section each week as a means of drawing the meeting to an early conclusion. No one will want, or be able, to discuss schedules and budgets after they hear your reading. If you are a teacher, read some of the sections aloud in class, either as inspiration or as real-life cautionary tales.

Do not read the entire book at one sitting, as it would render you too jazzed or too jaded to sleep for the next several days.

Save the final section (the one about Busch and Starling's "Rectal Foreign Bodies") for last, especially if you plan to read it aloud to a group.

Each of the Ig Nobel Prize winners has a much deeper and more intriguing story than it was possible to tell in this book. Use the references to find further information.

Go to the *Annals of Improbable Research* web site (www.improbable.com) for links to (in most cases) the winners' home pages, published work, and/or press clippings. There you will also find video of several of the Ig Nobel Prize ceremonies, and links to recordings of the annual Ig Nobel radio broadcast on National Public Radio's *Talk of the Nation/Science Friday with Ira Flatow* program.

We also publish, in the magazine (*AIR*) and in the free monthly e-mail newsletter (*mini-AIR*), news of the continuing adventures of past Ig Nobel Prize winners.

After you have read the book, you might find it

interesting to do two things. First, compare your impression of particular Ig winners with that of someone whose judgment you think you agree with. The question "Which of these are commendable and which damnable?" may reveal unexpected differences of opinion and personality.

Second, peruse the appendix that lists the winners year by year. Pick any year. Muse for a few moments about what ideas may have been batted around when that crop of winners met each other at the Ig Nobel Ceremony and playfully talked about combining their work. The discussion at the 1999 ceremony, for one, was particularly inspired.

A final word before you begin: THESE PEOPLE AND THEIR ACHIEVEMENTS ARE REAL. If you have trouble believing that – and you will – then use the references and GO LOOK IT UP YOURSELF TO SEE.

Medical Breakthroughs

The human body is always falling apart. Doctors and medical personnel labor mightily to stave off decrepitude or repair what is broken, infected, or just haywire. Sometimes it works, sometimes it doesn't. Sometimes it leads to an Ig Nobel Prize. Here are four of the medical achievements that have been so honored:

- Failure of Electric Shock Treatment for Rattlesnake Envenomation
- Transmission of Gonorrhea Through an Inflatable Doll
- Nose Picking in Adolescents
- Elevator Music Prevents the Common Cold

Failure of Electric Shock Treatment for Rattlesnake Envenomation

"The use of high-voltage electric shock therapy for the treatment of snake venom poisoning has recently gained popularity in the United States. We present a case that documents the dangerous, ineffective application of electric shock to the face of a patient envenomated by a Great Basin rattlesnake (Crotalus viridis lutosus)."

–from Dart and Gustafson's published report

THE OFFICIAL CITATION

THE IG NOBEL MEDICINE PRIZE

This prize is awarded in two parts. First, to Patient X, formerly of the US Marine Corps, valiant victim of a venomous bite from his pet rattlesnake, for his determined use of electroshock therapy: at his own insistence, automobile spark-plug wires were attached to his lip, and the car engine revved to 3,000 rpm for five minutes. Second, to Dr Richard C. Dart of the Rocky Mountain Poison Center and Dr Richard A. Gustafson of the University of Arizona Health Sciences Center, for their well-grounded medical report: "Failure of Electric Shock Treatment for Rattlesnake Envenomation."

Their report was published in *Annals of Emergency Medicine*, vol. 20, no. 6, June, 1991, pp. 659–61.

A former US Marine received a lesson on the theme "don't believe everything you read." The lesson involved his pet rattlesnake, a car, a too-cooperative friend, an ambulance, a helicopter, several liters of intravenous isotonic fluids, a battery of medications, and numerous medical personnel.

The man in question will be identified here as he is in the published medical report: "Patient X." Having already been bitten some 14 times by his poisonous pet snake,

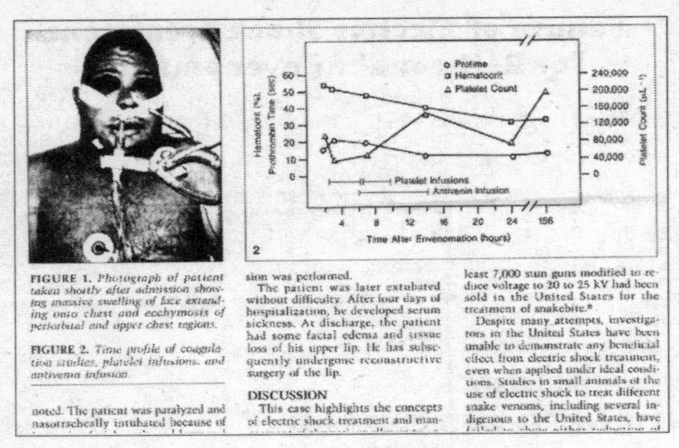

Protime
Hematocrit
Platelet Count
Hematocrit (%), Prothrombin Time (sec)
Platelet Count (μL⁻¹)
Platelet Infusions
Antivenin Infusion
Time After Envenomation (hours)
2

FIGURE 1. *Photograph of patient taken shortly after admission showing massive swelling of face extending onto chest and ecchymosis of periorbital and upper chest regions.*

FIGURE 2. *Time profile of coagulation studies, platelet infusions, and antivenin infusion.*

noted. The patient was paralyzed and nasotracheally intubated because of

sion was performed.

The patient was later extubated without difficulty. After four days of hospitalization, he developed serum sickness. At discharge, the patient had some facial edema and tissue loss of his upper lip. He has subsequently undergone reconstructive surgery of the lip.

DISCUSSION

This case highlights the concepts of electric shock treatment and man-

least 7,000 stun guns modified to reduce voltage to 20 to 25 kV had been sold in the United States for the treatment of snakebite.*

Despite many attempts, investigators in the United States have been unable to demonstrate any beneficial effect from electric shock treatment, even when applied under ideal conditions. Studies in small animals of the use of electric shock to treat different snake venoms, including several indigenous to the United States, have

The Prize-winning report.

Patient X did his best, he thought, to take precautions against a possible unlucky 15th chomp.

Though rattlesnake bites can be deadly, there is a standard treatment – injection with a substance called "antivenin." This almost always works, provided that the patient gets a sufficient amount soon after the bite occurs. For reasons that may at one time have been clear to him, Patient X was intent on using an alternative treatment.

He had read accounts in men's magazines of a powerful alternative treatment: application of a good, strong electric shock. High voltage was said to be essential. Some pundits recommended using an electric stun gun, and at least one company offered stun guns specially optimized for the purpose. Patient X and his friend agreed that, in the future, should either of them suffer a rattlesnake bite, the other would spring to the rescue with a bracing dose of electricity.

This ounce of prevention was worse than a pound of cure. Quite a bit worse.

One day, whilst Patient X was playing with his snake, the serpent embedded its fangs into Patient X's upper lip.

Patient X's friend immediately sprang into action. As per their agreement, he laid Patient X on the ground next to an automobile. He then connected Patient X to the car's electrical system, affixing a spark-plug wire to the stricken man's lip with a small metal clip.

The friend then revved the car engine to 3,000 rpm. To ensure a sufficient dose of electricity, he maintained that level for five minutes. As described in the medical report that was eventually published:

"The patient lost consciousness with the first electrical charge. An ambulance arrived approximately 15 minutes later to find the patient unconscious and incontinent of stool."

The ambulance attendants summoned a helicopter. During the flight, Patient X, rousing himself to some level of consciousness, fought off efforts to treat him.

A photograph taken shortly after he arrived at the hospital shows "massive swelling of face extending onto chest and ecchymosis of periorbital and upper chest regions." The man resembled a too-well-baked potato.

Dr Richard Dart and Dr Richard Gustafson, both then at the Arizona Poison and Drug Information Center at the University of Arizona Health Sciences Center, in Tucson, were brought onto the case. The treatment was complex and lengthy.

Of their patient's initial choice of treatment, Drs Dart and Gustafson commented that:

"Despite many attempts, investigators in the United States have been unable to demonstrate any beneficial effect from electric shock treatment, even when applied under ideal conditions … In addition, this treatment may have adverse effects."

Eventually, with a substantial amount of medical help and despite his own earnest efforts, the patient made a full recovery. Dr Dart and Dr Gustafson wrote an instructive technical account of the case, which they published in the *Annals of Emergency Medicine*.

For educating the public about treatments for snakebite, Patient X and the two doctors who saved his life shared the 1994 Ig Nobel Prize in the field of Medicine.

The winners could not travel to the Ig Nobel Prize Ceremony, but Dr Dart sent a tape-recorded acceptance speech. In accepting the Prize he said:

"I was stunned to receive this prize, although not as stunned as our patient."

The Dart/Gustafson medical report changed the way public health officials behave in regions where rattlesnakes abound. Public information campaigns now routinely include one extra item on their list of "Do Nots." An advisory from the Oklahoma Poison Control Center is typical. The list concludes with these items:

- **Do not** waste time capturing or killing the snake. Identification is helpful but not necessary.
- **Do not** apply a tourniquet.
- **Do not** pack wound in ice or apply heat.
- **Do not** give the victim a sedative or alcohol.
- **Do not** use a stun gun or electric shots.

Transmission of Gonorrhea Through an Inflatable Doll

"GONORESMITTET AV DUKKE"

–Front page headline of the Norwegian newspaper *VG*, the day after Ellen Kleist and Harald Moi were awarded the Ig Nobel Prize

THE OFFICIAL CITATION

THE IG NOBEL PUBLIC HEALTH PRIZE WAS AWARDED TO

Ellen Kleist of Nuuk, Greenland and Harald Moi of Oslo, Norway, for their cautionary medical report "Transmission of Gonorrhea Through an Inflatable Doll."

Their study was published in *Genitourinary Medicine*, vol. 69, no. 4, August, 1993, p. 322. (Note: The journal has since changed its name, and is now called *Sexually Transmitted Infections*.)

Dr Harold Moi decided to leave his native Norway for a few years and take a job at the Venereaklinikken – the venereal diseases clinic – in Nuuk, Greenland. At the time, neither he nor the clinic's nurse, Ellen Kleist, had had much experience with inflatable dolls.

Yet, when a taciturn seaman appeared at their clinic showing signs of disease and concealing all trace of an explanation, Doctor Moi and Nurse Kleist rose to the challenge. Using good medical detective work, they discovered the existence of the doll as well as its prominent role in giving their patient the clap.

The patient was captain of a trawler. He arrived at the Venereaklinikken with symptoms characteristic of gonnorhea. A blood test confirmed the diagnosis. Dr Moi and Nurse Kleist were puzzled, though, by the man's extremely sketchy story. They pressed him for details.

Their interrogation was not driven by curiosity alone. In Greenland, as in many parts of the world, medical professionals are required to help investigate any case of gonorrhea and track down its source, so as to prevent epidemics. There are similar requirements for certain other sexually transmitted diseases.

The skipper had been at sea for three months, but his symptoms had begun near the end of the voyage. Clearly, he had contracted gonorrhea on the boat. The question was how. The boat's crew were all male, and the skipper insisted that he had never had homosexual contacts.

Bowing to Dr Moi and Nurse Kleist's stern questioning, the skipper at last told his story. One night he had gone to a crewman's cabin and without permission borrowed a piece of equipment – the inflatable doll. A few days later he noticed the first symptoms of disease.

The owner of the doll had made good use of his property shortly before the skipper borrowed it. He (the doll's owner) was later tested for gonorrhea, and proved to have a raging case, courtesy of a live, uninflated girl he had visited just before going to sea.

The solution to this detective story was unique in the annals of medicine. Dr Moi and Nurse Kleist decided to write an account and send it a medical journal.

For their medical and literary triumphs, Harold Moi and Ellen Kleist were awarded the 1996 Ig Nobel Prize in the field of Public Health.

Dr Moi travelled to the Ig Nobel Prize Ceremony at his own expense. In accepting the Prize, he said:

"Ladies and gentlemen, I am honored and delighted to accept the famous Ig Nobel Prize. The biggest problem in this case was how to perform the mandatory partner notification and treatment. No reference in the literature to the pharmaco-kinetics of antibiotics in dolls could be found. So what else could be done than just give a shot and puncture?"

The following day, Dr Moi gave a lecture at Harvard Medical School before a rapt, standing-room-only audience of doctors. Immediately after the lecture, as the nearly wild applause died down, a Harvard professor in the front row told a colleague, "I learned something today."

Several weeks later, the Ig Nobel Board of Governors received a letter from a New York City resident who had read a newspaper account of Dr Moi's work. "This case," said the letter, "is a reminder to all of us that when you sleep with an inflatable doll, you are sleeping with everyone who ever slept with that inflatable doll."

IG NOBEL PRIZES

INFLATABLE DOLLS IN MODERN MEDICINE

In modern medical and nursing school curricula, inflatable dolls receive, at best, a fleeting mention. It is a rare medical textbook that devotes more than a few pages to the topic.

The medical literature does contain some intriguing research reports – "Use of Anatomical Dolls by Boston-Area Professionals" (1992), "Baby Dolls in Dementia" (1990), "Are the Genitalia of Anatomical Dolls Distorted?" (1990), the provocative "Anatomically Correct Dolls: Research vs. Clinical Practice" (1988), and the disturbing opthalmological report "A Maneuver to Elicit Vertical Dolls' Eye Movements" (1979). But none of these would have prepared Doctor Moi and Nurse Kleist for the case they had to handle.

Nose Picking in Adolescents

"BACKGROUND: Rhinotillexomania is a recent term coined to describe compulsive nose picking. There is little world literature on nose-picking behavior in the general population.

METHOD: We studied nose-picking behavior in a sample of 200 adolescents from 4 urban schools.

RESULTS: Almost the entire sample admitted to nose picking, with a median frequency of 4 times per day; the frequency was > 20 times per day in 7.6% of the sample. Nearly 17% of subjects considered that they had a serious nose-picking problem."

–from the published report by Andrade and Srihari

THE OFFICIAL CITATION

THE IG NOBEL PUBLIC HEALTH PRIZE WAS AWARDED TO

Chittaranjan Andrade and B.S. Srihari of the National Institute of Mental Health and Neurosciences, Bangalore, India, for their probing medical discovery that nose picking is a common activity among adolescents.

Their report was published as "A Preliminary Survey of Rhinotillexomania in an Adolescent Sample," *Journal of Clinical Psychiatry*, vol. 62, no. 6, June, 2001, pp. 426–31.

As the 21st century arrived, two distinguished psychiatrists offered mankind proof – written proof – that most teenagers pick their noses.

Dr Chittaranjan Andrade and Dr B.S. Srihari, colleagues at the National Institute of Mental Health and Neurosciences in Bangalore, India, were inspired by an earlier published report by scientists in the American state of Wisconsin. The Wisconsin research claimed that more than 90% of adults are active nose pickers, but it was silent on the question of whether teenagers are less picky, as picky, or more picky than their elders.

Dr Andrade and Dr Srihari decided to find out. They had a serious purpose. Virtually any human activity, if carried to excess, can be considered a psychiatric disorder, and nose picking is no exception. "While nose-picking behaviour in general appears to be a common and normal habit," they wrote, "it is necessary to determine the extent to which rhinotillexomania amounting to a disorder exists in the adolescent population."

Public Health Prize winner Chitteranjan Andrade at the 2001 Ig Nobel Prize Ceremony. Photo: Diana Kudarayova/AIR

They prepared themselves by reading other medical reports about nose picking. With few exceptions, those reports dealt with spectacular individual nose pickers, most of whom were psychotic. Dr Andrade and Dr Srihari learned that nose picking, as practiced by disturbed individuals, can be chronic, violent, and associated with nose bleeds. The two psychiatrists studied Gigliotti and Waring's 1968 report, "Self-Inflicted Destruction of Nose And Palate: Report of Case." They scoured Akhtar and Hastings' 1978 report, "Life-Threatening Self-Mutilation of the Nose." They marveled at Tarachow's 1966 report, "Coprophagia and Allied Phenomena," noting from it that "persons do eat nasal debris, and find it tasty, too."

Those cases all had their points of interest, but they could serve only as background material for the work Drs Andrade and Srihari had in mind. To determine the nose picking who, what, where, when, why, and how of a community, one must statistically sample the picking practices of many individuals.

Sampling is what the Wisconsin researchers did with adults. Sampling is what Drs Andrade and Srihari knew they must do with adolescents.

They prepared a written survey which included the questions opposite. (You might enjoy taking this survey yourself, or applying it to friends and colleagues.)

For their careful, scholarly, and compulsively humane approach to the study of nose picking, Chittaranjan Andrade and B.S. Srihari were awarded the 2001 Ig Nobel Prize in the field of Public Health.

Dr Andrade travelled from Bangalore, India, to Cambridge, Massachusetts, at his own expense, to attend the Ceremony. In accepting the Prize, he said:

"On behalf of myself and on behalf of everybody else who is happy for me today, I'm happy to accept this year's Ig Nobel Prize in public health. My work was on ... you won't believe it, just hold your breath – rhinotillexomania, which is a very fancy way of saying compulsive nose picking.

• In your opinion, what percentage of persons
in the population pick their noses?
• On average, how often in a day do you pick your nose?
• Do you sometimes pick your nose in public? (please answer YES or NO)
• Why do you pick your nose? (please tick as many as are applicable to you)
    • To unclog your nasal passages
    • To relieve discomfort or itch
    • For cosmetic reasons
    • For personal hygiene
    • Out of habit
    • For pleasure
• How do you pick your nose? (please tick as many as are applicable to you)
    • Using your fingers
    • Using an object such as tweezers
    • Using an object such as a pencil
• Do you occasionally eat the nasal matter that you have picked? (please answer YES or NO)
• Do you consider that you have a serious nose-picking problem? (please answer YES or NO)

Some 200 students answered the survey.
The results showed some surprising things.

• Nose-picking practices are the same for all social classes.
• Less than 4% of the students claimed they never pick their noses. Half of the students pick their noses four or more times a day. About 7% say they indulge 20 or more times a day.
• 80% use their fingers exclusively. The rest are split almost evenly in their use of tools, some choosing tweezers while others prefer pencils.
• More than half said they do it to unclog nasal passages or relieve discomfort or itching. About 11% claimed they do it for cosmetic reasons, and a similar number do it just for pleasure.
• 4.5% said they ate the nasal debris.

These figures are just highlights. The survey produced a wealth of data.

"Now, as you all know, having been adolescents yourself at some time, you've done things which were habitual, and I hope you haven't done things that were psychiatrically habitual such as trichotillomania, which means compulsive pulling of the hair, onychophagia, which means compulsive nail-biting, or rhinotillexomania.

"Some people poke their nose into other people's business. I made it my business to poke my business into other people's noses. Thank you, folks."

Two days later, Dr Andrade gave a public lecture and demonstration at the Ig Informal Lectures, elucidating the finer points of his research. In response to several questions, he assured anxious audience members that nose picking, in moderation, is "perfectly normal."

The *Times of India*, that nation's most prominent newspaper, reported the news on its front page with the headline "Ig Nobel for Indian Scientists Who Dig Deep."

Elevator Music Prevents the Common Cold

"The present study suggests at least the potential for an interesting new mode of defense against both the onset and the course of various disease entities (e.g., the common cold)."

–from a preliminary report distributed by Charnetski, Brennan, and Harrison

THE OFFICIAL CITATION

THE IG NOBEL MEDICINE PRIZE WAS AWARDED TO

Carl J. Charnetski and Francis X. Brennan, Jr, of Wilkes University, and James F. Harrison of Muzak Ltd in Seattle, Washington, for their discovery that listening to elevator Muzak stimulates immunoglobulin A (IgA) production, and thus may help prevent the common cold.

Their research report was published, a year after they won the Ig Nobel Prize, as "Effect of Music and Auditory Stimuli on Secretory Immunoglobulin A (IgA)," *Perceptual and Motor Skills*, vol. 87, no. 3, part 2, December, 1998, pp. 1163–70.

Can music juice up your immune system? Can frequent sex? Several years ago, psychology professor Carl Charnetski attended a meeting where he heard someone mention a chemical called "Immunoglobulin A." Professor Charnetski immediately began an ambitious research program which, so far, has involved Immunoglobulin A, music, journalists, sex, and the spit of many persons.

Prophetically for Professor Charnetski, Immunoglobulin A is also called "IgA." This chemical is one of many different so-called "antibodies" which the human immune system produces in response to infections or other dangers. Professor Charnetski reasoned that if he could find some common, pleasurable activity that causes the body to produce more of this chemical, he would have discovered an almost magical key to good health.

He and fellow professor Francis Brennan started looking for pleasurable activities that might have this effect. It would be easy to recognize, because someone's Immunoglobulin A level is easy to measure – a saliva test is all it takes.

The first pleasurable activity they tested: listening to music. The research was simple. They had volunteers listen to music, and spit.

In the earliest experiments, they had college students listen to musical notes. The students heard 30 minutes of upbeat, cheery notes. Then they heard 30 minutes of down-beat, melancholy notes. The cheery music produced higher Immunoglobulin A levels in the students' spit, but the dreary music produced lower levels.

Professors Charnetski and Brennan found this encouraging. Next, they teamed up with James Harrison of Muzak Ltd – the company that produces much of the world's elevator music – to do an experiment using more familiar kinds of music.

They tested four groups of people:

• One group listened to a 30-minute tape recording of so-called "environmental music," the kind of music some people call "smooth jazz."

• Another group listened to that same kind of music, but played from a radio instead of a tape.

• The third group listened to a half-hour of tones and clicking sounds.

• The fourth group was, in the researchers' words, "subjected to 30 minutes of silence."

The researchers tested everybody's spit.

Those who listened to smooth jazz from a tape recording had increased Immunoglobulin A levels in their spit – but those who listened to it on the radio did not.

The tones-and-clicks listeners' spit contained *decreased* levels of Immunoglobulin A.

Those who were "subjected to silence," like those who listened to smooth jazz on the radio, had unchanged spit.

Charnetski, Brennan, and Harrison announced that these findings were "significant" and could usher in a new era in the prevention of illness. For their harmonized attack on the common cold, Carl J. Charnetski, Francis X. Brennan, Jr, and James F. Harrison were awarded the 1997 Ig Nobel Prize in Medicine.

After mulling it over for a while, the winners decided that they could not, or would not, attend the Ig Nobel Prize Ceremony.

The team's research activities continued apace, although Harrison quietly dropped out of the picture.

Professors Charnetski and Brennan next explored how music affects the spit of newspaper reporters. That research was conducted on ten journalists in the newsroom of the Wilkes-Barre *Times Leader*. The results were encouraging, or at least suggestive, though perhaps not conclusive. (Full details are in the report "Stress and Immune System Function in a Newspaper's Newsroom," which appeared in the journal *Psychological Reports*.)

At that point, Professors Charnetski and Brennan switched their focus from music to sex. In 1999, they announced that college students who engage in frequent sexual intercourse have stronger immune systems than those who mate less frequently.

Two years later, they summarized all of their research in a book called *Feeling Good is Good for You*. The publisher's promotional blurb sums it up nicely:

"The media love to report how sex, laughter, and other simple pleasures are good for you. And you love to hear it.

But is inciting pleasure a legitimate medical prescription for boosting a person's immunity? Can you literally fight off infection with a smile? Researchers Carl Charnetski and Francis Brennan say yes."

Psychology & Intelligence

Psychologists study how people behave. People behave in unexpected ways. Sometimes, so do the psychologists studying them. Here are three examples:

- A Scholarly Study of Glee
- A Forbidding Experiment: Spitting, Chewing Gum and Pigeons
- Ignorance is Bliss

A Scholarly Study of Glee

"A phenomenon called group glee was studied in videotapes of 596 formal lessons in a preschool. This was characterized by joyful screaming, laughing, and intense physical acts which occurred in simultaneous bursts or which spread in a contagious fashion from one child to another. A variety of precipitating factors were identified... Group glee tended to occur most often in large groups (7–9 children) and in groups containing both sexes. The latter finding was related to Darwin's theory of differentiating vocal signals in animals and man."

–from Lawrence W. Sherman's published report

THE OFFICIAL CITATION

THE IG NOBEL PSYCHOLOGY PRIZE WAS AWARDED TO

Lawrence W. Sherman of Miami University, Ohio, for his influential research report "An Ecological Study of Glee in Small Groups of Preschool Children."

His study was published in *Child Development*, vol. 46, no. 1, March, 1975, pp. 53–61.

Lawrence W. Sherman was the first scientist to rigorously and systematically study, document, and analyze the occurrence of glee.

When Lawrence Sherman studied glee, why did he choose to study it in small groups of preschool children? Because that is where one finds the most gleeful glee.

Why did he choose to study glee? Because other psychologists hadn't, and because he needed to chose something as the subject of his PhD thesis, and because he himself is a gleeful person.

Sherman spent two years videotaping groups of three- and four-year-old children at their nursery school. Then he studied the tapes, classifying, describing, and probing the glee recorded therein.

He established a formal, scholarly definition of "group glee." Group glee is "a very intense, joyfully affective state maintained throughout a majority of the group (one half or more)."

Laypersons may be surprised to learn that there is a technical aspect of glee: it is a parameter called BEHAVIORAL MANIFESTATION.

Technically speaking, BEHAVIORAL MANIFESTATION is a complex consisting of three categories of overt behaviors through which group glee manifests itself, these categories being LAUGHTER, SCREAMING, and INTENSE PHYSICAL ACTS. These three behaviors can manifest either by themselves, or simultaneously in various combinations. The six fundamental combinations are:

(1) LAUGHTER
(2) SCREAMING
(3) LAUGHTER + SCREAMING
(4) LAUGHTER + INTENSE PHYSICAL ACTS
(5) SCREAMING + INTENSE PHYSICAL ACTS
(6) LAUGHTER + SCREAMING + INTENSE PHYSICAL ACTS

Sherman identified several key questions that any serious researcher must ask about an incidence of glee. Among them:

1. Was the glee disruptive?
2. Was the glee contagious? (That is, did it spread "from one person to another or throughout the entire group in a somewhat linear fashion, as in a chain reaction"?)
3. Was the glee – rather than being contagious – simultaneous? (That is, did "the children seem to get the signal or

input all at the same time so that the effect of the glee was like an explosion"?)

4. What was the duration of the glee? (Sherman timed each incident of glee "from the first overt signs until it ceased.")

Sherman identified 14 different things that can trigger glee. These include:

- A question from the teacher, such as "Who wants to go out and get John?"
- Incongruities or "funny words," such as "ya, ya, poo, poo, tat, tat."
- The breaking of a taboo, perhaps through the mention of taboo words. Sherman gives as examples "stinky-poo" and "shit."
- Someone else's misfortune – "A child who tripped over a milk carton," for example.
- Nothing at all. A considerable amount of glee comes from nothing at all.

Finally, Sherman performed a statistical analysis of glee. This turned up some notable facts.

Glee occurs approximately four out of every ten times children get together. Glee is generally short-lived – typically just four to nine seconds.

The most common expression of glee is joyful screaming without laughter; the least common is joyful screaming *with* laughter. Simple questions such as "Who wants to go out and get John?" are the most common triggers for glee.

Other people's misfortune or ineptness, we now know scientifically, is only rarely a cause for young children's glee. And we also know, thanks to Lawrence Sherman, that there is more glee when boys and girls are together than when they are apart.

For studying glee in an era of Valium, Lawrence W. Sherman won the 2001 Ig Nobel Prize in the field of Psychology.

Professor Sherman attended the 2001 Ig Nobel Prize Ceremony, where he was mobbed by a small group of gleeful children. In accepting the Prize, he said:

"I'm really impressed that this is happening at the end of my career. And I'm really pleased it didn't happen at the beginning."

Professor Sherman could not stay at Harvard for the subsequent several days of Ig Nobel-related events, because he had to return home for a meeting of the Ohio Gourd Association, of which he is vice president. As he was leaving to go to the airport, an admirer asked him "What is it you like about gourds?" Sherman's gleeful reply: "Everything!"

A Forbidding Experiment: Spitting, Chewing Gum, and Pigeons

"When I was a student in England, I was very impressed. Very orderly city, very courteous, very polite, very honest. Newspapers on sale on the roads outside the Tube station. Nobody there, just a box and coins and notes. You take it, you put your coins there. That's really a First World city. We try to be like that."

–Lee Kuan Yew, in a December 18, 1999 interview on Japan's NHK Television

THE OFFICIAL CITATION

THE IG NOBEL PSYCHOLOGY PRIZE WAS AWARDED TO

Lee Kuan Yew, former Prime Minister of Singapore, practitioner of the psychology of negative reinforcement, for his 30-year study of the effects of punishing three million citizens of Singapore whenever they spat, chewed gum, or fed pigeons.

While professional psychologists struggle to conduct their experiments on small groups of people, one dedicated, if uncredentialed, amateur psychologist tries out his theories on four million people at a time.

The populace of an entire country – Singapore – was banned from spitting, chewing gum, or feeding pigeons. These are the highlights of a many-fronted campaign to change people's trivial behaviors, and to do it in the way earlier psychologists trained rats, relying on punishment rather than persuasion.

The prohibitions were conceived and mandated by Lee Kuan Yew, the nation's former prime minister. Lee proudly, publicly, stated: "I think that a country has greater need of discipline than democracy." In Singapore, the discipline now covers a wide range of behavior.

Lee explained to a reporter that most Singaporean spitters are (like Lee himself) of Chinese descent:

"You know, the Chinese they spit everywhere. If you go to China, you can see them. And we started this very early. We say this is no good. This is a Third World habit. You spread tuberculosis, you spread all kinds of bad germs and diseases. So, we started with the school children, educated them, mass media and got the message home to the parents. And then we fine people. After they have been educated, they still do it, we fine them. And slowly, it has subsided."

On the global stage, Lee's anti-spitting campaign is not unprecedented, but it has two special twists.

In the late 19th and early 20th centuries, the United States and other nations waged campaigns against spitting. These were part of a broader effort to halt the spread of tuberculosis, and for the most part relied on public relations, not legal sanctions. Lee Kuan Yew's Singapore sputum law has a little to do with public health, but a very lot to do with propriety. Jet-setters do not spit. And where other nations had announced that spitting is a crying shame, Lee Kuan Yew's Singapore decreed that spitting is a shaming crime.

In January 1992, the Singaporean government banned the manufacture, import, or sale of chewing gum. Officially, the new law was a matter of public tidiness. Journalists, though, discovered that it was actually a response to a single incident in which someone used a wad of chewing gum to jam the door sensor of a subway car.

To Lee Kuan Yew and his government, pigeons are as undesirable, as abominable, as rats. The Singapore Housing and Development Board give the official explanation:

"Pigeons and crows visit the town because of the food

left by residents. Birds can bring with them certain health risks and nuisance such as food poisoning. In addition, the droppings of birds can soil the laundry, cars, walls and floors, and damage roof tiles."

Throughout Singapore, pigeon feeding was deemed not just undesirable, but impermissible.

In the aforementioned interview with the Japanese television network, Lee explained that he has been trying to "make this Third World country into a First World oasis." Buildings and roads, he said, are the straightforward part of that. But "[it's] very difficult, to change Third World habits into First World habits. So, it has to be through a long process of education."

In yanking his country's economy and social structure forward, Lee Kuan Yew has dared to tinker with conventional educational theory. In other parts of the world, education is not yet so straightforwardly based on fining, jailing, or (as with many Singapore laws) beating people.

As Prime Minister, and then in his role of power behind the throne, Lee has mounted many campaigns, against spitting, gum-chewing and pigeon-feeding, yes – but also against littering, smoking, and foul language. And he has pressed hard in favor of some of his favorite things: smiling, being courteous, and the scrupulous flushing of public toilets.

For his commanding research on how people ought to behave, Lee Kuan Yew won the 1994 Ig Nobel Prize in the field of Psychology.

The winner could not, or would not, attend the Ig Nobel Prize Ceremony.

Ignorance Is Bliss

"We argue that when people are incompetent in the strategies they adopt to achieve success and satisfaction, they suffer a dual burden: not only do they reach erroneous conclusions and make unfortunate choices, but their incompetence robs them of the ability to realize it.

"In 1995, McArthur Wheeler walked into two Pittsburgh banks and robbed them in broad daylight, with no visible attempt at disguise. He was arrested later that night, less than an hour after videotapes of him taken from surveillance cameras were broadcast on the 11 o'clock news. When police later showed him the surveillance tapes, Mr Wheeler stared in incredulity. 'But I wore the juice,' he mumbled. Apparently, Mr Wheeler was under the impression that rubbing one's face with lemon juice rendered it invisible to videotape cameras."

–from Dunning and Kruger's report

THE OFFICIAL CITATION

THE IG NOBEL PSYCHOLOGY PRIZE WAS AWARDED TO

David Dunning of Cornell University and Justin Kruger of the University of Illinois, for their modest report, "Unskilled and Unaware of It: How Difficulties in Recognizing One's Own Incompetence Lead to Inflated Self-Assessments."

Their study was published in the *Journal of Personality and Social Psychology*, vol. 77, no. 6, December, 1999, pp. 1121–34.

Everyone is incompetent, one way or another. David Dunning and Justin Kruger collected scientific evidence that incompetence is bliss.

Dunning and Kruger wanted to explore the breadth and depth of human incompetence. They staged a series of experiments at Cornell University, involving several groups of people. Before commencing the experimentation, they made some predictions, most notably:

1. That incompetent people dramatically overestimate their ability;
2. That incompetent people are not good at recognizing incompetence – their own or anyone else's.

In one experiment, Dunning and Kruger tested people's ability to tell whether jokes are funny – specifically, their ability to tell whether *other* people would laugh at the jokes.

They prepared a list of jokes. These jokes spanned a range from the officially Not Very Funny ("Question: What is big as a man, but weighs nothing? Answer: His shadow.") to the officially Very Funny ("If a kid asks where rain comes from, I think a cute thing to tell him is 'God is crying.' And if he asks why God is crying, another cute thing to tell him is 'probably because of something you did.'")

Dunning and Kruger then asked 65 test subjects to rate the funniness of each joke. They showed the same jokes to a panel of eight professional comedians – people who, as Dunning and Kruger point out, "make their living by recognizing what is funny and reporting it to their audiences." They then compared each test subject's ratings of the jokes with those of the professional comedians.

Some people had a very poor sense of what others find funny – but most of those same individuals believed themselves to be very good at it.

Dunning and Kruger realized that a sense of humor can be a tricky thing to judge, so their next experiment used tests that are easier to measure: logic questions from law-school entrance exams. The logic questions produced much the same results as jokes. Those with poor reasoning skills tended to believe they were the intellectual peers of Bertrand Russell or Mr Spock.

Overall, the results showed that incompetence is even worse than it appears to be. Not only do incompetent people not recognize their own incompetence; they also don't recognize competence when they see it in other people.

David Dunning explained why he took up this kind of research: "I am interested in why people tend to have overly favorable and objectively indefensible views of their own abilities, talents, and moral character. For example, a full 94% of college professors state that they do 'above average' work, although it is statistically impossible for virtually everybody to be above average."

Unskilled and Unaware of It: How Difficulties in Recognizing One's Own Incompetence Lead to Inflated Self-Assessments

Justin Kruger and David Dunning
Cornell University

People tend to hold overly favorable views of their abilities in many social and intellectual domains. The authors suggest that this overestimation occurs, in part, because people who are unskilled in these domains suffer a dual burden: Not only do these people reach erroneous conclusions and make unfortunate choices, but their incompetence robs them of the metacognitive ability to realize it. Across 4 studies, the authors found that participants scoring in the bottom quartile on tests of humor, grammar, and logic grossly overestimated their test performance and ability. Although their test scores put them in the 12th percentile, they estimated themselves to be in the 62nd. Several analyses linked this miscalibration to deficits in metacognitive skill, or the capacity to distinguish accuracy from error. Paradoxically, improving the skills of participants, and thus increasing their metacognitive competence, helped them recognize the limitations of their abilities.

The Prize-winning report.

Dunning and Kruger are themselves college professors (though at the time they did the experiment, Kruger was still Dunning's student). When they published their final report, the concluding words showed a degree of modesty: "To the extent this article is imperfect, it is not a sin we have committed knowingly."

For celebrating incompetence and unawareness, David Dunning and Justin Kruger won the 2000 Ig Nobel Prize in the field of Psychology.

The winners could not, or would not – or at least did not

– attend the Ig Nobel Prize Ceremony. It was and is unclear whether their absence was intentional.

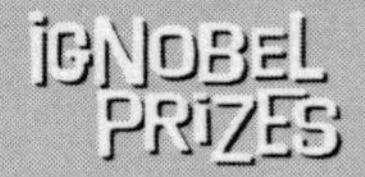

A LOVELY GIFT

If you have colleagues who are incompetent and unaware of it, Dunning and Kruger's research is a useful and convenient tool.

We recommend that you make photocopies of this report, and send them – anonymously, if need be – to each of those individuals. Repeat as necessary.

Economics

Almost everyone wants more money. Almost no one is quite sure how to get it.

Certain persons came up with startling economic insights which they apparently felt compelled to act upon. These people produced some memorable Ig Nobel Prize-winning achievements. Among them:

- The Economy of Chile versus J.P. Davila
- Squeezing Orange County/Bringing Down Barings
- The Good Lloyd's Shepherds Insure Disaster
- Dying to Save Taxes

The Economy of Chile versus J.P. Davila

"I feel obliged to introduce you to a new verb, davilar. To davilar, dear reader, is to commit a total, no-holds-barred, computer-generated stuff-up. The word springs from Santiago, Chile, where a few days ago the eponymous Juan Pablo Davila was being grilled by the carabineros, after certain unfortunate events."

–Columnist Charles Wright, in the Australian magazine *The Age*

THE OFFICIAL CITATION

THE IG NOBEL ECONOMICS PRIZE WAS AWARDED TO

Juan Pablo Davila of Chile, tireless trader of financial futures and former employee of the state-owned Codelco Company, for instructing his computer to "buy" when he meant "sell," and subsequently attempting to recoup his losses by making increasingly unprofitable trades that ultimately lost 0.5% of Chile's gross national product. Davila's relentless achievement inspired his countrymen to coin a new verb: "davilar," meaning, "to botch things up royally."

Juan Pablo Davila lost some money that belonged to his employer. By Davila's account, it was a matter of punching the wrong buttons on his computer terminal, and then making a panicked and miserably unfortunate attempt to fix things.

Before long, a humongous chunk of the Chilean national economy had disappeared, and the nation and the world were treated to a multi-continent fiesta of criminal proceedings and lawsuits.

Juan Pablo Davila was a not-very-senior employee of a Chilean government-owned company called Codelco. His job involved buying and selling contracts for minerals futures, mostly on the London Metal Exchange. With skill and luck, one could make huge profits as the price of copper, gold, silver, lead and other materials went up and down. But, beginning in late 1993, Davila fell into a losing streak – a very, very bad losing streak.

Outside his native country, almost no one heard of Davila until February 12, 1994, when *The Economist* reported what then appeared to be the story:

"Juan Pablo Davila claims that last September he made a mistake. He punched several "sell" figures into his computer as "buy," and vice versa. Mr Davila was a fairly junior executive at Codelco, Chile's mammoth state-owned copper company. But he handled all Codelco's minerals futures contracts. By the time he noticed his slip, he had already lost $40m. So he kept on dealing; when his credit lines finally ran out in January, his losses had reached $207m."

That amount – $207 million – was equivalent to about 0.5% of the entire Chilean gross national product.

A newspaper account in March 1994 described Davila as "a conscientious, but harassed, 34-year-old who apparently existed principally on cigarettes and *cafe negro*." The bemused tone showed up in fewer and fewer news reports, though, as word spread that fraud, rather than bumble-fingeredness, might be the story's underlying theme.

Fewer and fewer people regarded Davila as a hapless idiot. Eventually the government charged him with illegal trading activities. Allegedly, he had been working not only for Codelco but also for a rival, privately owned Chilean copper company, booking his unprofitable trades for Codelco as profitable trades for the other firm. The press began reporting that Davila had also taken huge kickbacks from other firms, among them London-based Sogemin Metals Ltd and Germany's Metallgesellschaft AG.

In subsequent months, Davila's name seemed to acquire a two-word prefix – most news reports called him "rogue trader Juan Pablo Davila." In Chile the name "Davila" became an everyday word: "davilar." At first, this new verb meant "to screw up royally and very expensively." But, in some mouths, it left an aftertaste of ripe, sharp, cheesy scheming.

For his ever-downward, ever-widening spiral of losses, Juan Pablo Davila won the 1994 Ig Nobel Prize in the field of Economics.

The winner could not, or would not, attend the Ig Nobel Prize Ceremony. He was busy attending to legal matters.

Davila's lawyer blamed it all on Codelco's senior managers. "They authorized these futures operations," he told the Chilean newspaper *La Epoca*. "It is incomprehensible that they were not controlled. It's like saying to someone 'go and play this money at the horse races.' In this case they said 'go play with the salary of Chile,' which is copper, 'at the race tracks,' where Davila could win or lose."

Davila's tangled affairs tied up courts in Chile, England, and the United States, and involved an impressive array of companies and individuals.

In 1997, Davila began serving a three-year prison term for tax evasion. He managed to gain early release, and has since tried to keep a low public profile.

Squeezing Orange County/ Bringing Down Barings

"My sincere apologies for the predicament I have left you in."

–Nick Leeson, in the resignation letter he faxed to Barings Bank

THE OFFICIAL CITATION

THE IG NOBEL ECONOMICS PRIZE WAS

AWARDED JOINTLY TO Nick Leeson and his superiors at Barings Bank and to Robert Citron of Orange County, California, for using the calculus of derivatives to demonstrate that every financial institution has its limits.

For an introduction to the work and legacy of Nick Leeson, see the book *Rogue Trader: How I Brought Down Barings Bank and Shook the Financial World*, by Nick Leeson, Little, Brown, 1996.

For an introduction to the work and legacy of Robert Citron, see the book *Big Bets Gone Bad: Derivatives and Bankruptcy in Orange County*, by Philippe Jorion, Academic Press, 1995.

1. Risk can be profitable!
2. Risk can be exciting!
3. Risk can be risky!
...
...
86. Risk can be disastrous.

This succession of thoughts, or one much like it, may have occurred to both Robert Citron and Nick Leeson as each whiled away his time in prison. Each man had taken a series of risky gambles with other people's money, and found disaster beyond his wildest dreams.

These were two scandals of nearly mythic import, occurring one on the heels of the other. Thanks to Citron, one of America's wealthiest counties (were it a nation, Orange County would have been the 30th largest in the world) suddenly went bust. Thanks to Leeson, one of England's oldest banks suddenly went bankrupt.

Robert Citron and Nick Leeson made bold investments, buying and selling things called "derivatives."

What is a derivative? Well, the definition wasn't so important – it appears that neither Citron nor Leeson really understood what a derivative is. What mattered was that each man had tremendous self-confidence – and tremendous self-confidence is the hallmark of the born financial genius. Both men had been hailed as true geniuses, because for a while each had brought staggering success.

Robert Citron was the treasurer of Orange County, California. He invested (or as some would later say, "gambled") the county's money in stocks and derivatives. At first, he was very skilled (or lucky), and made fantastically huge profits.

Nick Leeson was a trader in the Singapore office of Barings Bank, one of Britain's most venerable institutions. He "invested" the bank's money in stocks and derivatives. At first he was very "skilled" and made fantastically huge profits.

In October, 1994, Citron's investments all went completely sour. Orange County went bankrupt.

In February, 1995, Leeson's investments all went completely sour. Barings Bank collapsed.

After things blew up, supervisors at Barings and in Orange County expressed great surprise. The financial press delighted in describing these parallel catastrophes. A 1996 report from *Bloomberg Business News* put it pithily:

"Regulators who have picked up the pieces from similar disasters – from Nick Leeson's $1.4 billion of losses for Barings PLC to Robert Citron's $1.7 billion of losses for Orange County, Calif. – say the fallout generally follows a familiar pattern: the institution blames a lone trader. As more evidence develops, it becomes clear the losses stem as much from bosses who were willing to overlook trading risks as from the deceptions of an individual.

"'Nobody calls you a rogue trader until you start losing money,' said Philip McBride Johnson, a Washington attorney who was chairman of the Commodities Futures Trading Commission from 1981 to 1983. 'It's amazing how people can do pretty much what they want as long as they make money.'

"Orange County supervisors, who were required by law to help oversee the county's finances, said they didn't understand Citron's investment strategy or the risks involved. Peter Norris, head of Barings's investment banking, said that none of the company's top managers actually understood the intricacies of derivatives trading."

Just days before Orange County filed for bankruptcy, Citron was forced to resign his post.

Just hours before Barings crashed, Leeson fled his office in Singapore. His first stop was Malaysia, then it was on to Brunei, Thailand, and finally Germany, where the police gave him accommodation in a lovely jail cell. After six months of spirited negotiation, Leeson was whisked back to Singapore for a rendezvous with the criminal justice system there.

Orange County went into bankruptcy. The remains of Barings Bank were sold, for the price of £1, to the Dutch banking and insurance company ING.

For their accomplishments, Robert Citron and Nick Leeson shared the 1995 Ig Nobel Prize in the field of Economics.

The winners could not, or would not, attend the Ig Nobel Prize Ceremony. Each had a previous engagement.

Citron had begun serving a five-year prison term (eventually reduced to one year). He revealed that in deciding where and how to invest Orange County's money, he had consulted not just Merrill Lynch and other large corporate financial advisers, but also a local psychic and a mail-order astrologer.

Nick Leeson wrote a book about his adventures.

Leeson had begun serving a six-year stretch (eventually reduced to two years) in Tanah Merah prison, Changi, Singapore. While there he co-authored a charming book called *Rogue Trader: How I Brought Down Barings Bank and Shook the Financial World*. The book ends as he is about to be extradited from Germany to Singapore. Leeson muses fondly about his former supervisors:

"I realized that I was glad to have played my part in this fiasco rather than theirs. I was happier in my prison cell than they were, sitting at home nursing their credibility back to pieces and always knowing what their friends were saying behind their backs. Fuck 'em! I thought."

Since emerging from prison, Leeson has gone on to the lecture circuit where, according to press accounts, he is paid as much as $100,000 to warn audiences about the need for tighter corporate controls and regulation.

The Good Lloyd's Shepherds Insure Disaster

"Whenever life has seemed a little glum in the past year or two, I have been able to console myself with the reflection that I am not a member of Lloyd's."

–Max Hastings, editor of the *Daily Telegraph*,
as quoted in the book *Ultimate Risk*

THE OFFICIAL CITATION

THE IG NOBEL ECONOMICS PRIZE WAS AWARDED TO

The investors of Lloyd's of London, heirs to 300 years of dull, prudent management, for their bold attempt to insure disaster by refusing to pay for their company's losses.

Several books try to describe the Lloyd's of London story.
One is *Ultimate Risk: The Inside Story of the Lloyd's Catastrophe*,
by Adam Raphael, Four Walls Eight Windows, 1995.

Over the span of three centuries, Lloyd's of London became the biggest, most innovative, most influential, most respected, and most profitable insurance company in the world. Then, just like that, it was coming apart at the seams.

Lloyd's investors are obligated by law and sacred oath to personally pony up for the company's losses. During the first 300 years, Lloyd's almost always made a profit, but the moment that changed, most of those investors – including many of great wealth, power, and fame – refused to pay up. And that put the company into one hell of a spot.

A lot of people made a lot of very bad decisions.

Lloyd's of London has always been peculiarly organized, and, until the 1980s, it was peculiarly profitable. In its later years, Lloyd's became a mess, the kind of mess that is awkwardly draped atop a larger mess, a larger mess that is somehow supported by a more massive mess, a massive mess which manages to conceal within it a type of mess that beggars all description.

Until it blew up, Lloyd's of London was entirely owned by a select group of individuals known as the "Names." To be a "Name" was considered a privilege and an honor; only the wealthy and powerful were called. Unlike investors in other companies, the Names pledged to cede Lloyd's of London their own fortunes in the then-unlikely event that the company incurred losses. Happily, Lloyd's almost never incurred losses. Nearly every year brought nice fat profits, which the company parceled out to the little gaggle of Names.

'Twas a happy and seemingly perpetual arrangement. And then things changed. Rapidly. A series of natural disasters – Hurricane Betsy, the Exxon Valdez oil tanker spill, etc. – caused high losses. Many of the Names got angry, and threatened not to pay up.

Lloyd's managers scrambled to add new investors – lots of 'em. They loosened the company rules so that just about anyone, not just the rich, could become a Name. Suddenly Lloyd's had thousands of new Names. Many were Americans and Canadians who pledged their right arms – and their entire net worth, no matter how tiny – for the chance to (maybe, just maybe!) join the British upper crust. Where, in 1970, there had been 6,000 Names, most of them reasonably wealthy, by 1987, there were 30,000, many of them lower middle class.

When the company then incurred massive losses – £500 million in 1988, £2 billion in 1989, almost £3 billion the next year, many of the new Names were asked to fork out everything they owned. For them it had been "Welcome to Lloyd's, Welcome to Instant Bankruptcy!"

The situation was grim. Most refused to pay. Instead, they filed lawsuits.

But wait – there was more. Lloyd's management had

done something clever. They'd gotten the government to exempt Lloyd's from many of the nation's basic financial rules, making it difficult for the Names to sue.

But wait – there was more. Under Lloyd's bizarre accounting rules, the ruined Names owed the company not just everything they owned at the moment, but also much of their earnings in the future. And no one could say how much they would still owe in the future. No one could even say if there would ever be a limit to the amount. And if the ruined Names were to die, the obligations would then fall upon their children.

And was there more? Of course there was. In the cozy, traditional Lloyd's way of arranging things, many of the older, wealthier Names and some of the better-connected new Names were *not* asked to pay anything.

More? Glad you asked. Many of the older Names resented the pleadings of those who were ruined, and made a big point of not helping them. *Noblesse* not *oblige*.

For the whole sad, twisted mess, the Names of Lloyd's of London won the 1992 Ig Nobel Prize in the field of Economics.

The winners could not, or would not, attend the Ig Nobel Prize Ceremony.

New lawsuits popped up all over the place. The list of Names shrivelled – by the year 2001, there were fewer than 3,000 individuals left. Lloyd's did manage to keep its institutional nostrils above water; they did it by selling stock to corporate investors, savvy businesses that refused to accept the romantic pledge-your-every-last-iota requirement. As to the future: no one can say for how long Lloyd's famous bell will continue to toll, or for whom.

Dying to Save Taxes

"On January 15, 2000, the New York Times reported that in the first week of the new millennium local hospitals had recorded an astonishing 50.8% more deaths than in the last week of 1999. The Times suggested that this phenomenon was due to infirm people willing themselves to stay alive long enough to witness the dawning of the new age. Apparently, the anticipation of momentous events can motivate people to live longer."

–from the economics report "Dying to Save Taxes..."

THE OFFICIAL CITATION

THE IG NOBEL ECONOMICS PRIZE WAS AWARDED TO

Joel Slemrod, of the University of Michigan Business School, and Wojciech Kopczuk, of the University of British Columbia, for their conclusion that people find a way to postpone their deaths if that would qualify them for a lower rate on the inheritance tax.

Their report was published as "Dying to Save Taxes: Evidence from Estate Tax Returns on the Death Elasticity," National Bureau of Economic Research Working Paper No. W8158, March, 2001.

By doing some intense, skillful detective work, Joel Slemrod and Wojciech Kopczuk found evidence that people will do pretty much anything for money – even die for it.

Economists like to believe that people make rational decisions and base all their actions on cool, self-interested thought. In the backs of their minds, though, they wonder.

Joel Slemrod wondered quite a bit, and he asked a simple question that few economists had dared to raise:

"Could the timing of death be, to some extent, a *rational* decision? Economists presume that the timing of other

Ig Nobel Prizewinner Joel Slemrod prepares to give his acceptance speech. Ig Nobel Minordomo Julia Lunetta gently buffs him. Photo: Jon Chase/Harvard News Office.

important events, such as childbearing or marriage, may be so affected – why not dying itself?"

Being a professor of Business Economics and Public Policy, and Director of the Office of Tax Policy Research at the University of Michigan, Slemrod knew how to seek the answer. Together with his prize graduate student Wojciech Kopczuk, he sifted through nearly a century's worth of tax records.

Other economists had dug into the slightly-less-existential questions of whether people time their marriage ceremonies to take advantage of tax laws, or whether they time the conception and birth of their children to maximize the tax benefits. "If birth," Slemrod and Kopczuk asked, "why not death?"

The idea of timing one's exit had been looked into, at

least slightly, by physicians. Medical libraries are full of reports analyzing when and how people tiptoe off life's stage. (One called "Postponement of Death Until Symbolically Meaningful Occasions," published in 1990 in the *Journal of the American Medical Association*, claims that the death rate "dips before a symbolically meaningful occasion" such as a big religious holiday, "and peaks just afterward.")

Many countries impose a tax on those who inherit substantial possessions. It goes under different names – inheritance tax, estate tax, death tax, etc. Exactly what is taxed and at what rate varies considerably from nation to nation, in some cases from locality to locality, and in many cases from one year to another.

In the United States, where Slemrod and Kopczuk live and work, the first such tax of real consequence was instituted in 1916. Under various political pressures, the rate bounced up and down quite often. Slemrod and Kopczuk looked at what happened when the estate tax rate increased markedly on eight occasions (twice in 1917, and once each in 1924, 1932, 1934, 1935, 1940, and 1941) and decreased on five others (in 1919, 1926, 1942, 1983, and 1984).

The analysis was complex, but it all boiled down to a simple conclusion:

"There is abundant evidence that some people will themselves to survive in order to live through a momentous event. Evidence from estate tax returns suggests that some people will themselves to survive a bit longer if it will enrich their heirs."

Slemrod and Kopczuk are modest about their work. "To be sure," they say, "the evidence is not overwhelming." They also mention the possibility that, in some cases, relatives intentionally report an incorrect date of departure.

For their work, Joel Slemrod and Wojciech Kopczuk won the 2001 Ig Nobel Prize in the field of Economics.

Joel Slemrod traveled to the Ig Nobel Prize Ceremony, at his own expense. In accepting the Prize, he said:

"Well, never in my wildest dreams did I think it would be like this. I'm pleased to accept the Ig Nobel for myself and my co-author, who I think is watching the video stream up in Vancouver. Also, I think my son and daughter are watching. Hi, kids. We are pleased to accept this award because we believe in the spirit of the Ig, that the pursuit of science, even social science, can be fun and that sometimes we learn by pushing our hypotheses into extreme and unlikely places. Our research provides evidence about something everybody already knows, that some people will do anything for money. Of course, other people live their whole lives with no regard for money, and sorting this out is a continuing challenge for economics.

"Little did we know that when we did this research, the US Congress, in its wisdom, would vote to abolish the US estate tax for the year 2010 – and *only* the year 2010 – setting up the best natural experiment for a hypothesis ever conceived. Somebody, I think maybe Benjamin Franklin, once said that the only two inevitable things are death and taxes. Well, come 2010, it will be death *or* taxes."

Peace – Diplomacy & Persuasion

Masters of diplomacy and peacemaking are, in some cases, highly idiosyncratic. Here are four outstanding examples:

- Parliamentary Punching and Kicking
- The Levitating Crime Fighters
- Daryl Gates, the Gandhi of Los Angeles
- Stalin World

Parliamentary Punching and Kicking

"Initially, Chien and Lin were merely engaged in verbal sparring over these accusations, after which many law makers left the room. Chien then requested the chairman call for order. It was at this moment that Lin approached Chien and accused him of disrupting the meeting. Lin then grabbed Chien by the neck and the two wrestled amid efforts by other law makers to separate them. Lo also joined in the melee.

"After the groups had been separated, Lin and Lo demanded that Chien apologize for what they said were his groundless accusations against them. When Chien left the room, Lin and Lo pursued him. The incident up to this point had been recorded live by TV cameras and broadcast nationwide.

"What the cameras did not show, however, was what occurred after the three left the conference room. According to Chien, Lin punched him on the chin. Both Lin and Lo deny this."

–report in the *Taipei Times*, January 4, 2000

THE OFFICIAL CITATION

THE IG NOBEL PEACE PRIZE WAS AWARDED TO

The Taiwan National Parliament, for demonstrating that politicians gain more by punching, kicking and gouging each other than by waging war against other nations.

Democracy at its best can be rough-and-tumble. What can rival the unbridled passion, drama, and intellectual thrill of legislators grappling with ideas, wielding words as weapons, and ultimately beating concepts into laws?

In 1988, some new members of the Taiwan legislature decided to find out.

Dr Ju Gau-jeng claims to have personally originated the new style of legislating in 1988, when he was a member of the Democratic Progressive Party ("the DPP"). Ju leaped onto a table, and started a donnybrook.

This was not long after Taiwan's election laws loosened up to allow opposition parties a shot at office. Until then, opposition had been banned or squashed for the 40 years since General Chiang Kai-shek led his glorious retreat from the mainland and founded a new government on the island of Taiwan.

Under Taiwan's constitution, Parliament was packed with legislators who represented districts back on the mainland, districts they had not seen in decades and with which they had no contact. With few responsibilities, these now-elderly gentlemen divided their time between snoozing and saying "no."

The newly elected opposition legislators found this frustrating. After Dr Ju Gau-jeng took his famous first steps onto the table and into parliamentary prize-fighting history, others soon joined in.

Wang Chao-wei says he was the next innovator. Wang overturned seven tables laden with food at a luncheon honoring the elders.

Ju, Wang, and a number of their fellow DPP members worked hard over the next few years to convince their septuagenarian and octogenarian opponents to retire. They used the sweet science of persuasion. They pushed. They punched. They kicked. They clobbered. Parliament members were bloodied and bowed. Bandages, crutches and neck braces made regular appearances on the parliamentary fashion scene.

Finally, in 1991, the old men from the imaginary districts had had enough, and voted themselves out of office.

With their favorite opponents gone, the gentlemen legislators from the DPP saw no reason to stop exercising their skills. The DPP was still the minority party, with few weapons to use against the ruling party. And so they continued to use their charm, directing it now more broadly against anyone who, from their point of view, declined to engage in rational discourse.

The tactics were a popular hit, and soon the Taiwan city council was staging fist fights of their own. The stately National Senate joined in, too. One of the first brawls there started with a remark about one female senator's underwear. The remark led to a slap, which escalated to just short of a pie-in-the-face free-for-all.

For Taiwan's law-making bodies, the tradition has continued ever since.

For striking the first blows and then going the distance, the Taiwan National Parliament won the 1995 Ig Nobel Peace Prize.

The winners could not, or would not, attend the Ig Nobel Prize Ceremony.

They received a second, more informal, honor two years later, when the Dalai Lama, Tibet's exiled spiritual leader and winner of the Nobel Peace Prize, visited Taiwan. The Dalai Lama succinctly praised Parliament for its spirit and style. His exact words: "It's okay!"

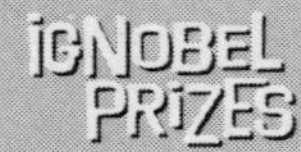

BEST OF THE BOUTS

Some of the parliamentary fights are classics. Here are press accounts of three of them.

FROM *ASIA WEEK*, JUNE 16, 1995
It was not your average tea break. Rivals in Taiwan's parliament, angry over a procedural issue, splashed tea on each other and threw punches during a brawl that erupted during the reading of a bill. Careful to steer clear of the fight were members of the Democratic Progessive Party, who are actively seeking to change their reputation as political street fighters.

FROM THE *CHINA POST*, MARCH 29, 2001
Independent Legislator Lo Fu-chu yesterday angrily slapped his People First Party colleague Diane Lee (Lee Ching-an) for allegedly calling him a gangster, hospitalizing the woman with minor injuries. The scuffle took place at around 9 a.m., soon after Lo stepped into a Legislature room where Lee was expecting to begin a meeting of the education and culture committee.

According to security-camera footage of the scuffle, Lo began by arguing with a seated Lee across the table, as Lee smacked the tabletop, stood up and picked up a paper cup of water, apparently preparing to throw it at Lo.

In response, Lo quickly struck the cup from Lee and then ran around to Lee's side of the table in a bid to hit her, but was blocked by a male aide of Lee. Lee's aide was soon shoved away by an aide of Lo. The two lawmakers then began to fight with each other, during which Lo managed to slap Lee's shoulder, grab her hair, and punch her in the head, even though other lawmakers and police at the scene were trying to block him.

FROM THE AUSTRALIAN TELEVISION NEWS PROGRAM *60 MINUTES*, 1995
ANNOUNCER: He hit you...
MR WEI: He hit my head, and the blood...
(Footage of men fighting in parliament)
ANNOUNCER: (Voiceover) The dispute was over a budget vote for a new nuclear power plant. The combatants: Legislator Wang and the KMT's Hsui Thai Chung.
MR HSUI THAI CHUNG: (Through translator) And when he started pushing me, I–I took his head around that way and started pushing back the other way.
(Footage of men fighting in parliament)
MR WEI: Somebody hit him, you know. I don't know who.
ANNOUNCER: You didn't have anything to do with it.
MR WEI: I don't know who hit–hit–hit him on the neck. I don't know.
ANNOUNCER: Uh-huh.
MR WEI: But he said, you know, 'Somebody hit–hit me.'
MR CHUNG: (Through translator) So it's the–the fourth, fifth and sixth vertebrae...
ANNOUNCER: Were damaged?
MR CHUNG: Yeah.
(Through translator) Were seriously damaged.
ANNOUNCER: So it was very painful.
MR CHUNG: (Through translator) Forever, this is–this is a permanent injury.
ANNOUNCER: OK. So at the end–at the end of the day in the Taiwanese parliament, we had one man with a bloodied head – you.
MR WEI: Yes.
ANNOUNCER: And Mr Hsui, who ended up with a neck brace.
MR WEI: Yes.
ANNOUNCER: That's quite a day.
MR WEI: You know, pol–our politics are very difficult, you know.
ANNOUNCER: You can say that again.
MR WEI: In more–in more crises, very difficult.
ANNOUNCER: Yes, and painful.
MR WEI: In–in Taiwan–in Taiwan – and painful.
(Footage of men fighting in parliament; Hsui Chung)
ANNOUNCER: (Voiceover) The next day, the fight was joined by hundreds of constituents from Mr Wang's and Mr Hsui's electorates. It was a matter of honor, each side defending the pride of their battered local member.

The Levitating Crime Fighters

"BONN (Reuters) – Yogic flyers from Natural Law Parties around the world bounced for peace on Friday in Bonn pledging to combat global crime, disease, war and unemployment with meditation and levitation ...

"A group of 23 yogic flyers, clad in white trousers and T-shirts decorated with the party's rainbow symbol, sat cross-legged on a foam mat with their eyes closed. After a few minutes of meditation, the flyers began to shake, giggle and hop across the mat on their knees, bouncing about half a metre off the ground and often colliding with each other as they projected themselves forward.

"'Yogic flying is enormously beneficial for the individual producing maximum coherence in brain functioning,' said John Hagelin, the party's US presidential candidate in 1995 [sic], before settling down for his daily hop. Hagelin, a physicist by profession, said sociological studies had shown dramatic social improvements if the square root of 1% of the population practised transcendental mediatation and yogic flying every morning and evening ...

"Reinhard Borowitz, Secretary General of the Maharishi Council of Natural Law Parties, said the group wanted to set up specially trained teams of yogic flyers around the world to do away with the need for national armies and weapons."

–from a 1997 Reuters news report

THE OFFICIAL CITATION

THE IG NOBEL PEACE PRIZE WAS AWARDED TO

John Hagelin of Maharishi University and the Institute of Science, Technology and Public Policy, promulgator of peaceful thoughts, for his experimental conclusion that 4,000 trained meditators caused an 18% decrease in violent crime in Washington, DC.

His study was published as "Interim Report: Results of the National Demonstration Project To Reduce Violent Crime and Improve Governmental Effectiveness In Washington, DC, June 7 to July 30, 1993," Institute of Science, Technology and Public Policy, Fairfield, Iowa.

In June and July of 1993, a group of scientists performed a bold experiment.

Their aim: to drastically reduce the amount of violent crime in Washington, DC – a famous den of murder, rape, and robbery.

Their method: to scientifically and systematically blanket the city with the mental emanations from transcendental meditation and yogic flying.

John Hagelin is, by his own admission, a remarkable man. He is Professor of Physics and Director of the Institute of Science, Technology and Public Policy at the prestigious Maharishi University of Management, in Fairfield, Iowa. He is a practiced expert in quantum physics, transcendental meditation, yogic flying, and running for President of the United States.

John Hagelin is very concerned about crime.

"As a Dartmouth- and Harvard-trained, unified-field theoretical physicist," he wrote in a letter to a newspaper, "I have been fortunate to have worked closely with the world's foremost scientist in the field of consciousness, Maharishi Mahesh Yogi. As a patriot and scientist, I am prepared to provide our government with the scientific knowledge and the proven, natural-law-based solutions to the problems that confront the nation."

John Hagelin is very concerned about crime.

In 1992, he was the Natural Law Party's candidate for President of the US. He was not elected that year. He ran again in 1996 and 2000. Records indicate that he was not elected in either of those years. The Natural Law Party is based at the Institute of Science, Technology and Public Policy at the prestigious Maharishi University of Management, in Fairfield, Iowa, and has branches in England, Germany, India, Switzerland, Thailand,

Bermuda, Croatia, Latvia, Argentina, and some 70 other nations.

John Hagelin is very concerned about crime.

In 1993, he perfected a method for preventing violent crime.

In technical terms, the method consists of "forming coherence groups in major cities to lower the stress throughout our society, in order to alleviate the prime cause of criminality." In plain words: Hagelin pays people to meditate and levitate themselves off carpets. When a sufficient number of skilled people do this at the same time and in the same place, the crime rate drops. It's as simple as that.

He demonstrated this method in the summer of 1993. From June 7 through July 30, 4,000 trained meditators meditated and levitated themselves in and/or near Washington, DC.

At a press conference held a year later, just weeks before the presidential election, candidate John Hagelin announced the results of his experiment: it was a success. While the meditators were meditating and levitating, Washington's crime rate had dropped by 18%.

Technically speaking, that is. The rate of actual crimes committed in Washington did not drop by 18% – in fact, during the experiment, Washington's weekly murder rate hit the highest level ever recorded. However, the crime rate was 18% lower than what John Hagelin's computer predicted it would have been had not 4,000 trained meditators been meditating and levitating.

For his influence on criminals, John Hagelin won the 1994 Ig Nobel Peace Prize.

The winner could not, or would not, attend the Ig Nobel Prize Ceremony.

In subsequent years, John Hagelin continued his experiments. Early in 2001, Hagelin, together with the Maharishi Mahesh Yogi and Indian Major General Kulwant Singh, held a news conference in Washington, DC, to announce a fundraising drive. The purpose: to raise $1,000,000,000, the interest from which would pay for a squad of 40,000 trained yogic levitators, who would patrol war zones and thus bring peace to the world. They expressed confidence that they could persuade people to invest in their plan.

And in summer of 2002, Hagelin held a press conference to inform all parties in the Middle East that peace would soon be achieved there, once he and his trained meditators and levitators swung into action, which would happen as soon as they received the massive – but given the task at hand, modest – funding they desired.

Daryl Gates, the Gandhi of Los Angeles

"By now it was finally becoming clear to me that the King episode was having repercussions far beyond the unrestrained beating of a man."

–Los Angeles Police Chief Daryl Gates, in his book *Chief...*

THE OFFICIAL CITATION

THE IG NOBEL PEACE PRIZE WAS AWARDED TO

Daryl Gates, former Police Chief of the City of Los Angeles, for his uniquely compelling methods of bringing people together.

There are several books about the Rodney King riots and Chief Gates's role pertaining to them. One is Official Negligence: *How Rodney King and the Riots Changed Los Angeles and the LAPD*, by Lou Cannon, Times Books, 1997. Daryl Gates himself co-authored a book in which he to some degree addresses the subject: *Chief: My Life in the LAPD*, by Daryl F. Gates with Diane K. Shah, Bantam Books, 1992.

Daryl Gates ran the most publicized police force in history, an organization glorified in movies and books, and especially on television. Then, one day, TV stations began showing – and endlessly repeating – videotape of several police officers, all of whom were white, viciously beating a traffic violator who was black. The public demanded that the cops be punished. It was at this point that Chief Gates, using a charismatic blend of actions and pronouncements, inaction and silence, helped the small, nasty incident grow into something big and long-lived. In several parts of the city, aggrieved people came together and started rioting. Every shocking bit was televised, uniting the many peoples of the earth in voyeuristic revulsion.

In the wee hours of March 3, 1991, a group of Los Angeles police officers chased a very drunken man named Rodney King as he sped his car down a freeway. They caught him, and a week later, millions of television viewers watched videotape of the officers beating and beating and beating

Rodney King with metal batons, and also kicking and stomping on him. Los Angeles is the TV and movie capital of the universe and, by 1991, seemingly every Los Angelino owned a video camera. The man who taped the Rodney King beating was a neighbor who had been awakened by the sirens and the shouting, and who happened to have a brand-new camera he wanted to test out.

For decades, television shows such as *Dragnet*, *Columbo*, *Hunter*, and *Adam-12* broadcast attractive pictures of the Los Angeles Police Department (LAPD). The LA cops in these programs were always courteous and humane. The LA cops in the Rodney King video appeared to be quite something else.

The LAPD had always had a seamy, unpublicized side. Charges of police brutality – especially against black and Latino citizens – were so common that the city government expected to pay vast sums every year in legal settlements.

The Rodney King incident set the city's teeth on edge. Four of the cops were put on trial, charged with using excessive force. Having seen the videotape, the world expected them to be convicted of at least some of the charges. Everyone in Los Angeles feared, though, that if the jury somehow let the accused cops off the hook, rioting would erupt in the streets.

An all-white jury somehow *did* let the accused cops off the hook, and as the news spread across town, riots *did* erupt. These were the biggest riots the United States had seen in more than two decades.

For years, Chief Gates had been bragging that his police force, having learned from past failures, was fully trained and fully prepared to cool any potential disturbance the moment it started. It turned out he was wrong.

Under Chief Gates's direction, or lack of direction, the

LAPD was disorganized beyond disorganized and did almost nothing to quell the riots until it was far too late. Bizarrely, as people were being attacked and killed in the streets, and as buildings and cars were being torched, and as the world watched it all on television, Chief Gates left his post to go attend a political fundraising event. By the time he returned several hours later, things had gotten completely out of hand.

At the height of the troubles, Rodney King himself stepped back into the spotlight and played a curiously calming role. He went on television and plaintively asked "Can't we all get along?"

Two months after the riots ended, with much of the public screaming for his head, Daryl Gates resigned as police chief of the city of Los Angeles. Although he had certainly not caused the troubles, his gritty determination and tough-guy posturing had helped bring thousands of riotous Los Angelenos and millions of television viewers together in ways they had never imagined.

For this, Daryl Gates won the 1992 Ig Nobel Peace Prize. The winner could not, or would not, attend the Ig Nobel Prize Ceremony. The Ig Nobel Board of Governors arranged for Mr Stan Goldberg, manager of the Crimson Tech Camera Store in Cambridge, Massachusetts, to accept the Prize on behalf of Chief Gates. Here is the text of Mr Goldberg's acceptance speech:

"As general manager of Crimson Tech Camera Store, I am pleased to accept this award on behalf of Daryl Gates. Daryl Gates has done more for the videocamera industry than any other individual. He has shown the world how a good quality videocamera can capture the memories of a generation. [Here, Mr Goldberg held up a videocamera.] Take this baby, for instance. It's a model VHS-C with one-lux light sensitivity, and AF power zoom/macro lens, full-range autofocus, and automatic time-date stamping. We sell it for just $599.98, which includes a free bonus case. We'll beat any competitor's advertised price..." (At this point several people rushed on stage, attacked Mr Goldberg, and removed him from the building. It is believed that a member of the audience videotaped the incident and offered it for sale to the television networks.)

In July 1992, a very few months after resigning as police chief and just three months before being awarded the Ig Nobel Prize, Daryl Gates published a hastily written (well, co-written) autobiography, titling it *Chief: My Life in the LAPD.* The book's ending is as forthright and blunt as the man himself:

Soon after his resignation as police chief, Gates and a co-author rushed his autobiography into print.

"By the beginning of 1992, I knew it was time to leave. I was bored. After 14 years as chief, the challenges were gone; there wasn't anything I hadn't done ... I had stayed as long as I cared to. No one had run me out."

Daryl Gates went on to subsequent careers as a radio talk show host and a video game designer.

Stalin World

"You may have thought Disneyland and Stalin-era mass deportations had nothing in common. But thanks to enterprising Lithuanian Viliumas Malinauskas, they do now. The 60-year-old canned mushroom mogul recently opened an odd-ball park that mimics a Soviet prison camp. The facility – part amusement park, part open-air museum – is circled by barbed wire and guard towers, and dotted with some 65 bronze and granite statues of former Soviet leaders Vladimir Lenin and Josef Stalin, and communist VIPs. Organizers say it's the first and only Soviet theme park in the world. Officially, the 30-hectare complex is called the Soviet Sculpture Garden at Grutas Park. But residents of the nearby village of Grutas have dubbed it 'Stalin World' – a name that's stuck."

–from the magazine *City Paper – Baltic States*, 2001

THE OFFICIAL CITATION

THE IG NOBEL PEACE PRIZE WAS AWARDED TO

Viliumas Malinauskas of Grutas, Lithuania, for creating the amusement park known as "Stalin World."

The Grutas Sculpture Park, better known as
"Stalin World," is at Grutas, 4690 Druskininkai, Lithuania.
Telephone: (370 233) 55484, 52507, 52246, 47709
Fax: (370 233) 47451.

Viliumas Malinauskas is a man of immense strength, imagination, and mordant humor. Long ago, he was heavyweight wrestling champion of Lithuania. He served, not exactly by choice, in the Soviet army, and later managed a collective farm. After the communist system collapsed, Mr Malinauskas created an international mushroom distribution business, which made him wealthy. But he was somewhat bored until 1998, the year a collection of Soviet statues went up for auction. These were not just any statues. These were gigantic granite and bronze Lenins, Stalins, and other Official Soviet Heroes. Mr

Malinauskas realized that a man with money and wit, and a lack of affection for the old Soviet Union, might do something interesting with them.

A promotional flyer from the Grutas Sculpture Park, which is popularly known – for good reason – as "Stalin World".

They say of Stalin World that it "combines the charms of a Disneyland with the worst of the Soviet Gulag prison camp." Viliumas Malinauskas, on the whole, rather agrees. The neighbors are of varying opinions.

Stalin World is the only place in the neighborhood that's surrounded by barbed wire, and guard towers, and loudspeakers blaring Soviet-era military music.

A few people baulked at the idea of Stalin World, and tried to throw up roadblocks to its construction. Mr Malinauskas at that point was 58 years old, tougher than nails, and had spent most of his life taking on Soviet apparatchiks. These nosy anti-Parkers stood no chance against him.

Viliumas Malinauskas accepts the Ig Nobel Peace Prize, aided by the interpreter he brought with him from Lithuania to the ceremony at Harvard University. Four Nobel Laureates and numerous other dignitaries saluted Malinauskas by donning Stalin masks. Photo: Caroline Coffman/*Annals of Improbable Research*.

When Stalin World opened on April Fools' Day, 2001, Mr Malinauskas honored the most vocal of these gadflies by putting up wooden statues of them near the front entrance. "Talking to them is just like talking to these," Mr Malinauskas told a reporter from the *Sydney Morning Herald*, knocking the head of one of the wooden statues. "They have decided I'm wrong, but I say let the public be the judge."

To build the park itself, Mr Malinauskas drained a swamp, and installed more than a mile of winding wooden walkways. The ambiance is pleasant, with pine, birch, and fir trees, a children's play area with seesaws and swings, and a small petting zoo with exotic birds. There's a cozy little cafe, information aplenty, and a place to buy souvenirs.

But the statues are the main attraction. There are more

than 60 of them – Stalin, Lenin, and the other old favorites, larger and even harder than life. Some are missing body pieces – a head gone here, a hand or thumb there. But that just adds to the charm.

For visitors of a certain sensibility, the park's highlight is a replica of a 1941 railway station where guards in Soviet uniforms direct them onto cattle cars which shuttle to a reconstruction of a prison camp. This is not for everyone.

On opening day, actors dressed as Lenin and Stalin handed out shots of vodka and tin bowls of borscht. Mr Malinauskas wants the experience to be on the whole fun, and for those who are so inclined, also educational.

At the park's entrance, big red signs proclaim, "Happy New Year, Comrades!" The guides conduct tours in Lithuanian, Russian, or English. Overall, visitors find the experience to be much more pleasant than a place called Stalin World has any right to be.

For conceiving of Stalin World and making it a reality, Viliumas Malinauskas won the 2001 Ig Nobel Peace Prize.

Mr Malinauskas, accompanied by his wife and an interpreter, traveled at his own expense from Grutas, Lithuania to the Ig Nobel Prize Ceremony at Harvard. In accepting the Prize, this is what Mr Malinauskas's interpreter said he said:

"Mr Malinauskas wants to begin his speech with sincere greetings from a small carnival world that is situated near the Baltic Sea from Lithuania. He was invited here to tell you about the Grutas Park, which is popularly known here as Stalin World. According to such a given name, he would like to make you a present of a bronze bas-relief of Stalin.

"At the end of his speech, he would like to remind this honorable audience one very known old truth. It's better to see something once than to hear about it ten times, isn't it?

So, dear ladies and gentlemen, he invites all of you to visit the wonderful country of Lithuania and as well to see the Grutas Park."

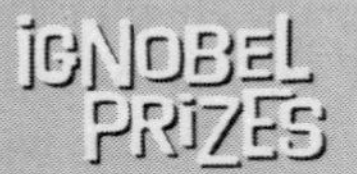

A POEM FOR STALIN WORLD

IN STALIN WORLD
by Alice Shirrell Kaswell
(with apologies to Colonel John McCrae, author of *In Flanders Fields*)

In Stalin World the statues grow
Lenin and Stalin, row on row;
And from the proletariat,
Feliks Dzerzhinsky, with a hat.
Maybe that's where the pigeons go.

They've fallen far, these metal ghosts,
And ended up as tourist hosts
In Stalin World.

Please come to Grutas! Come and see
These mighty men from history.
Check out their prices on the list,
And buy yourself a communist
From Stalin World.

Peaceful Ejaculations & Explosions

Some peacemakers favor an explosive approach to the process. Here are four exemplars:

- Auto Thieves Flambé
- Booming Voices of Britain
- Father of the Bomb
- Pacific Kaboom

Auto Thieves Flambé

"My personal feeling is that it would definitely blind a person – he will never see again. This is definitely nonlethal. A person won't just stand there and let you roast him."

– Charl Fourie

THE OFFICIAL CITATION

THE IG NOBEL PEACE PRIZE WAS AWARDED TO

Charl Fourie and Michelle Wong of Johannesburg, South Africa, for inventing an automobile burglar alarm consisting of a detection circuit and a flame-thrower.

The device is described in World Patent number WO9932331, "A Security System For A Vehicle" (1999); and also South Africa Patent ZA9811562, similarly titled, (2000).

When Johannesburg experienced a rising wave of car thefts and hijackings, a husband-and-wife team decided to do something about it. They would make these crimes too hot for criminals to go near.

Charl Fourie and Michelle Wong did what most good inventors do – combine existing kinds of technology into something wholly new, wholly greater than the sum of its parts. Automobile burglar alarms had been around for a while. So had flame-throwers. The marriage of the two, like the marriage of Charl Fourie and Michelle Wong, was obviously meant to be.

They gave their invention a simple name: "The Blaster." It can be installed in and under virtually any car or truck. The main components are gas canisters (the size of these can range from 6.6 to 19.8 pounds) feeding into a system of tubes. The gas tank is generally installed in the trunk. The pipes are mounted beneath the vehicle body, with their business ends aimed to the left and right, beneath the doors.

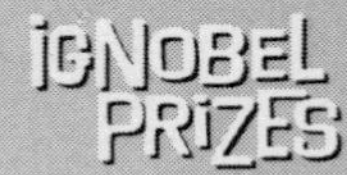

A PATENTED THIEF ROASTER

Here are selected highlights from the patent for "The Blaster."

FIELD OF THE INVENTION
This invention relates to a security system for a vehicle, more specifically to a security system which will assist in preventing vehicle hijacking.

BACKGROUND TO THE INVENTION
Vehicle hijackings are a serious problem in certain countries. A conventional modus operandi is for a hijacker to approach the vehicle from at least the driver's side and usually armed with a firearm. Conventional security devices fitted to vehicles are normally effective only once the hijacker has taken possession of the car and is sitting therein. Few security devices exist which disable the hijacker while still outside the car and in such a way that the hijacker is given little or no opportunity to shoot at the driver.

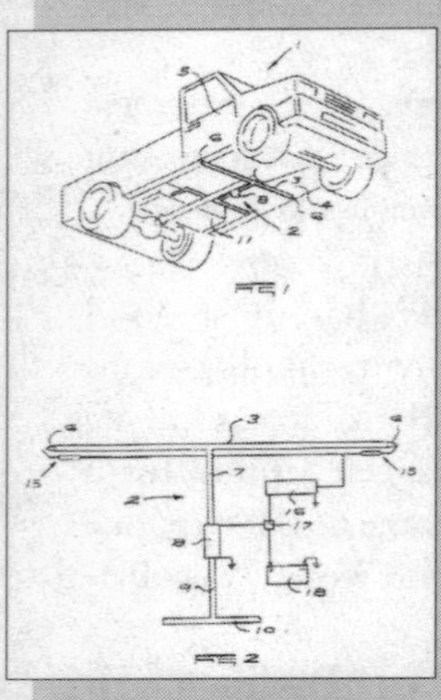

Technical drawings from the patent for The Blaster.

DETAILED DESCRIPTION
The switch (17) is located within the interior of the vehicle (1) to be operable by a foot of the driver (not shown) of the vehicle (1).

In use, when a hijacker (not shown) stands along one side of the vehicle (1), the driver depresses the switch (17) twice with his foot, the first time to place the switch (17) on standby and the second time to cause it to operate the pump (8) and coil (16). Hereafter, the pump (8) draws fuel from the fuel line (10) and projects it under pressure through the nozzles (6). The igniters (15) simultaneously create a spark at the outlet of each nozzle (6) to ignite the fuel being projected there through. This causes a large flame which continues while the switch (17) is depressed. The nozzles (6) are slightly upwardly directed so that the flame is projected towards the torso of the hijacker. Also, the nozzles produce a fairly widely dispersed spray, so that flame covers a fairly large area along the side of the vehicle.

It is envisaged that the security device will provide a highly effective method of disabling hijackers in the vicinity of the side of the vehicle while not causing much or any damage to the vehicle itself.

The driver operates a foot pedal, which ignites a blast of flame to both sides of the car simultaneously.

Fourie says that if people pull up at traffic lights, stop their car and then someone comes up to the window with a gun and tells them to get out, they should put their hands up and then step on the gas. "This is a case of opting for the lesser of two evils," he explained to the BBC.

The Johannesburg police head of crime intelligence, David Walkley, announced that he has a Blaster installed in his car. He told Reuters that "There is nothing that says this is illegal. It depends entirely on the circumstances and whether you can justify self-defense. Yes, there are certain risks in using it, but there are also risks in not having anything at all." Walkley later said that endorsements are contrary to police regulations, and that he was not endorsing the product.

Charl Fourie told reporters that "The demand is huge." This huge demand translated ultimately into total sales of about 200 Blasters.

For attempting to make the streets safer, Charl Fourie and Michelle Wong won the 1999 Ig Nobel Peace Prize.

The winners could not, or would not, attend the Ig Nobel Prize Ceremony.

The Blaster made it into the *Guinness Book of World Records* in the category of "Criminals" – "most dangerous car security device."

Commercially, the product did not completely fulfill its inventors' hopes. Auto manufacturers declined to make it standard equipment in new cars and, even as a do-it-yourself add-on, The Blaster never made it into the top-seller category.

Two years later, Fourie and Wong came out with a pocket-sized, hand-held version of the device, the details of which are published in South Africa Patent #ZA200001559, "A Hand Held Security Device."

Booming Voices of Britain

"This is part of the Armed Forces' continuing efforts to achieve the best possible value for money."

–A Royal Navy spokesman interviewed by the *Daily Telegraph*, May 20, 2000

THE OFFICIAL CITATION

THE IG NOBEL PEACE PRIZE WAS AWARDED TO

The British Royal Navy, for ordering its sailors to stop using live cannon shells, and instead just to shout "Bang!"

The British Royal Navy found a way to save money and also bring a measure of peace and quiet. The method, albeit more traditional than some people realized, was hailed as an innovative, and perhaps portentous, way to ring in the year 2000.

The story, in a nutshell, was told in the May 20, 2000 issue of the *Guardian*:

"The Royal Navy has barred trainees at its top gunnery school from firing live shells and ordered them to shout 'bang' in a cost-cutting exercise. Sailors at the Gunnery and Naval Military Training School on land-based HMS Cambridge, near Plymouth, Devon, have been told to cry out into a microphone rather than pull the trigger after they have loaded shells and aimed their guns. They have previously fired live rounds from land-based turrets. It is believed that the move to cut back on shells, which cost £642 each, could save the Ministry of Defence £5 million over three years.

"One sailor said: 'You sit in a gun and shout 'Bang, bang.' You don't fire any ammunition. It's a big joke and the sailors are disgusted.' The sailor, serving on a warship, added: 'Junior ratings are coming aboard and they cannot fire guns without specialists watching them. It is causing dismay. You used to hear the sound of gunfire coming from HMS *Cambridge* – now all there is are shouts of 'Bang, bang' over the microphone.'"

The imperative to say "Bang" is in fact an honored tradition in the British armed forces. In his book *Adolf Hitler: My Part in His Downfall*, military historian Spike Milligan documents an instance of Just-Say- "Bang"ism in which he took part during World War II:

"There was one drawback. No Ammunition. This didn't deter [our sergeant-major], he soon had all the gun crews shouting 'Bang' in unison. 'Helps keep morale up,' he told visiting [General] Alanbrooke. By luck a 9.2 shell was discovered in Woolwich Rotunda. An official application was made: in due course the shell arrived. A guard was mounted over it. The Mayor was invited to inspect it, the Mayoress was photographed alongside it with a V for Victory sign: I don't think she had the vaguest idea what it meant. A month later, application was made to HQ Southern Command to fire the shell. The date was set for July 2nd, 1940. The day prior, we went round Bexhill carrying placards: 'THE NOISE YOU WILL HEAR TOMORROW AT MIDDAY WILL BE THAT OF BEXHILL'S OWN CANNON. DO NOT BE AFRAID.'"

Milligan further reports that the shell turned out to be a dud.

That was the Army, and that was 1940. The Navy has always tried to be more methodical than its sister service, and thus sometimes works more slowly. It took them a full 60 years to adopt the new innovation, but adopt it they did.

For bravely taking firm, quiet action, the British Royal Navy won the Ig Nobel Peace Prize in the year 2000.

The winners could not, or would not, attend the Ig Nobel Prize Ceremony. Instead, Richard Roberts, a 1993

Nobel Laureate in Physiology or Medicine and himself a native of England, accepted temporary custody of the Prize. (See below.) Roberts also vowed to spend the next year, if necessary, seeking someone in the Navy to whom he could hand over the Prize. He did spend the year trying, but he did not succeed. Dr Roberts still has custody of the Prize, and it is hoped that some high official of the Navy will get in touch with him to arrange for delivery to a proper permanent place of enshrinement.

Nobel Laureate Richard Roberts accepts temporary custody of the Ig Nobel Prize on behalf of the winners. Photo: Andrea Kulosh/*Annals of Improbable Research*.

In the weeks following the Ceremony, the Ig Nobel Board of Governors received letters from angry citizens of other nations, especially Germany, pointing out that their own militaries also Just Said "Bang," and so deserved a share of the Royal Navy's Ig Nobel Prize.

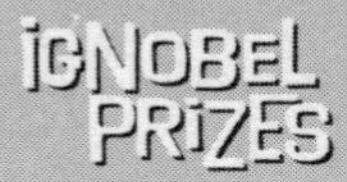

THIS IS THE SPEECH DELIVERED BY NOBEL LAUREATE RICHARD ROBERTS ACCEPTING TEMPORARY CUSTODY OF THE BRITISH ROYAL NAVY'S PRIZE.

"I must admit I never thought I'd be here representing the British Navy. Fortunately they abandoned conscription in England one year before I was eligible.

"However, I have to admit that it's very demeaning to have to stand there and say 'bang.' So I brought along an alternative which I'm hoping the Royal Navy will accept. [At this point Dr Roberts brandished a toy gun that sprouted a 'Bang' flag.]

"Now, in the interests of cutting costs, it occurred to me that there are a number of ways that the Navy might think about doing this, one of them related to these terribly expensive uniforms that they have to put on. In consultation with my 11-year-old daughter Amanda, she came up with an admiral's hat. [At this point, Dr Roberts donned a splendidly constructed, paper admiral's hat.]

"There's one other thing. You know, it's really awfully expensive to keep moving those boats around all the time. It occurred to me that perhaps one thing that could happen is we could replace most – maybe all – of the boats with some plastic boats that we just anchored in strategic places. Remember, this worked for the Royal Air Force during the War when they set up these dummy planes to fool the Germans in East Anglia. And now a moment of British pride: those of you who watched the Olympics know that the Brits did rather well in rowing. Suggests another alternative for the Navy."

Father of the Bomb

"He's a danger to all that is important. I really do think it would have been a better world without Teller."

–I.I. Rabi, winner of the 1944 Nobel Physics Prize and a senior colleague of Teller's on the Manhattan Project

THE OFFICIAL CITATION

THE IG NOBEL PEACE PRIZE WAS AWARDED TO

Edward Teller, father of the hydrogen bomb and first champion of the Star Wars weapons system, for his lifelong efforts to change the meaning of peace as we know it.

Edward Teller was one of the great scientists of the 20th century. Brilliant, gregarious, and blessed with the knowledge that he was always right, Teller also yearned to be one of the most influential scientists on world affairs. In personality and results, he was explosive.

Edward Teller is, in a sense, Mr Bomb. He was part of almost every technical and political step in the history of the first atomic bomb and its progeny. He helped persuade the US government to build that first one, and took part in the now-famous Manhattan Project at Los Alamos, New Mexico, which did the actual building. History books say he spent most of his three years at Los Alamos nagging people and dreaming about the future. His fondest dream was to plan a new, even more powerful, kind of bomb.

The first atomic bomb was based on nuclear fission – splitting atoms apart to release a huge amount of energy in an explosion. Teller wanted a bomb that would squeeze atoms so much they would fuse together, releasing far, far more energy in a much, much bigger explosion. This new device would be called a "thermonuclear bomb."

Teller was still fond of the old-fashioned atomic bomb –

but now he would use it in a supporting role. Just as it takes a little detonator cap to trigger off an old, old-fashioned chemical bomb, it takes a little atomic bomb to trigger off a thermonuclear bomb.

He pushed extremely hard to persuade the US government and the military to let him build a thermonuclear bomb. He got his way, though others did much of the technical work.

Many of the scientists who had built the earlier bomb urged caution. They took seriously a warning Teller himself gave them the first time he explained his plan. The thermonuclear bomb might generate enough heat to ignite the gases in the atmosphere or the oceans, effectively burning the earth's surface to a crisp. Teller didn't see this as a huge problem. Nor was he worried about other worries: the possible long-term radiation effects on bystanders; the possibility that building a new kind of weapon would cause other countries to bend heaven and earth to build one too; the astronomical and endless cost of continuing to do this particular kind of technical work, and of the arms race.

The new bomb got built and tested. The atmosphere did not ignite, nor did the oceans. The Soviet Union did race to catch up and build their own thermonuclear bomb, and they succeeded. The cost of developing, building, and maintaining the bombs did rise even higher than anyone had foreseen, and much of the world was frightened to death.

Edward Teller was delighted. He continued to push for new kinds of weapons, the more technically difficult and expensive the better. He turned his wonderful imagination to new kinds of missile systems for sending bombs to far-off places. He noted that the Soviets tried to match everything he and his admirers came up with, so he vowed to always stay a few steps ahead, no matter what the cost, and no matter whether it was possible to get the thing to work.

As the decades rolled on, many of the new weapons did not, in fact, work, but Teller was relentlessly inventive and even more relentlessly demanding. If he could conceive of a weapon, then so might an enemy, and that was good reason to spend money on it. If he could conceive of a weapon to use against other weapons that he'd conceived of, even better.

Most historians believe that the so-called "Star Wars" missile defense plan of the 1980s got taken seriously, and got lavishly funded, because Edward Teller pushed so hard for it. The plan itself reportedly popped out of the fertile technical mind of US President Ronald Reagan, who thought it would be a splendid idea, though he himself didn't know what the idea was.

That and the other Teller-backed projects all have exciting names – X-ray laser weapons, "Brilliant Pebbles" kinetic-energy space-based interceptors, "pop-up deployment," "Super Excalibur," the "High Frontier" initiative. The list is long, and the investment in them is endless. These weapons by themselves are not enough to prevent the people of the earth from blowing each other up, Teller cautions, but they are necessary first steps in building enough weapons to do the job.

For supplying the world with such bursts of enthusiasm, Edward Teller won the 1991 Ig Nobel Peace Prize.

The winner could not, or would not, attend the Ig Nobel Prize Ceremony.

Pacific Kaboom

"Chirac said that while many world leaders had condemned French nuclear testing publicly, few had criticized him privately ... Chirac singled out the Australian government for its 'excessive' reaction, saying: 'I'm not angry, I'm sorry. I don't see why they did that. It's demagogic.'"

–from a Reuters news report, October 23, 1995

THE OFFICIAL CITATION

THE IG NOBEL PEACE PRIZE WAS AWARDED TO

Jacques Chirac, President of France, for commemorating the 50th anniversary of Hiroshima with atomic bomb tests in the Pacific.

Right after he took office, Jacques Chirac ordered up a fireworks display that would make everyone respect the power and glory of France.

On May 17, 1995, Chirac was sworn in as president. On June 13, he announced that his country would set off a string of thermonuclear bombs – on the other side of the world – triumphantly ending its three-year moratorium on testing nuclear weapons.

They would need a bit of quiet time to prepare, he said, but other than that delay, nothing would stop the show.

On July 16, he quietly celebrated the 50th anniversary of the first atomic explosion at Alamogordo, New Mexico.

On August 6, he quietly celebrated the 50th anniversary of the dropping of an atomic bomb on Hiroshima.

On August 9, he quietly celebrated the 50th anniversary of the dropping of an atomic bomb on Nagasaki.

On August 10, President Chirac announced the capstone of his plans. France would spend an entire year or so performing its series of wonderful nuclear fireworks exhibitions. Then, after their final performance, they would retire from the business, and support an international Comprehensive Test Ban Treaty so that no one ever again would set off "any nuclear weapon test explosion or any other nuclear explosion."

On August 17 came an annoying distraction from the competition. China, the only nuclear power that hadn't stopped setting off test explosions, set off another of its homemade bombs at its Lop Nor test site. In France, some thought this a very strange thing – a country exploding a nuclear bomb on its own territory.

September 5 was Jacques Chirac's big day. France set off a 20-kiloton thermo-nuclear explosion at the Moruroa Atoll in the south Pacific.

New Zealand and Australia, which are much closer to what was left of Moruroa than France is, got angry and complained. President Chirac said they were being "demagogic." New Zealand even asked the International Court of Justice in The Hague to tell France to call off its other scheduled performances. On September 22, the court rejected New Zealand's request.

On October 1, France set off a 110-kiloton thermonuclear bomb at the Fangataufa Atoll, which is not far from Moruroa.

October 6, Jacques Chirac won the 1995 Ig Nobel Peace Prize.

The winner could not, or would not, attend the Ig Nobel Prize Ceremony.

He celebrated by having France set off a 60-kiloton thermonuclear bomb on October 27, back at Moruroa Atoll, then set off further thermonuclear fireworks displays at the same place on November 21 (40 kilotons) and December 27 (30 kilotons).

After a pause to celebrate New Year's, they staged a 120-

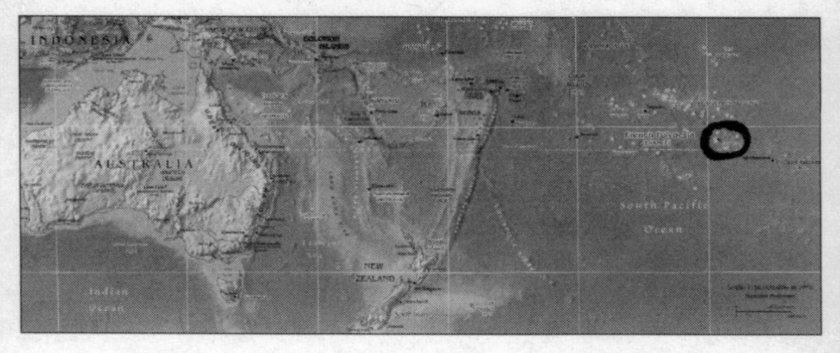

A map showing the Pacific neighborhood that includes Australia and New Zealand (bottom left) and Moruroa and Fangataufa Atolls (circled). Map courtesy of the US Central Intelligence Agency.

kiloton fireworks display on January 27, at Fangataufa. Two days later, Jacques Chirac called an early end to the exhibitions. The massive international protests had nothing to do with the early ending, President Chirac said. "I know that the decision that I made last June may have provoked, in France and abroad, anxiety and emotion. I know that nuclear weaponry may cause fear. But, in an always dangerous world, it acts for us as a weapon in the service of peace."

Love & Marriage

Every year millions of people obsess about love and marriage and, after marriage, many of them wonder about fidelity and infidelity. Singly and in combination, all these ideas have led to Ig Nobel Prize-winning achievements.

This chapter describes two of them:

- The Compulsive Biochemistry of Love
- Marriage By the Million

The Compulsive Biochemistry of Love

"The evolutionary consequences of love are so important that there must be some long-established biological process regulating it. Recent findings suggest that the serotonin (5-HT) transporter might be linked to both neuroticism and sexual behaviour as well as to obsessive-compulsive disorder (OCD). The similarities between an overvalued idea, such as that typical of subjects in the early phase of a love relationship, and obsession, prompted us to explore the possibility that the two conditions might share alterations at the level of the 5-HT transporter."

–from Marazziti, Rossi, Cassano, and Akiskal's report

THE OFFICIAL CITATION

THE IG NOBEL CHEMISTRY PRIZE WAS AWARDED TO

Donatella Marazziti, Alessandra Rossi, and Giovanni B. Cassano of the University of Pisa, and Hagop S. Akiskal of the University of California (San Diego), for their discovery that, biochemically, romantic love may be indistinguishable from having severe obsessive-compulsive disorder.

Their study was published as "Alteration of the Platelet Serotonin Transporter in Romantic Love," *Psychological Medicine*, vol. 29, no. 3, May, 1999. pp. 741–5.

Hundreds, perhaps thousands, of songs, poems, novels, and movies explore the link between obsession, compulsion, and romantic love. Donatella Marazziti, Alessandra Rossi, Giovanni B. Cassano and Hagop S. Akiskal undertook the first thorough biochemical investigation of this complex and delicate question.

Doctors Marazziti, Rossi, Cassano, and Akiskal did things, as all good scientists do, systematically. They began by honorably declaring their interest: "Since falling in love is a natural phenomenon with obvious implications for the process of evolution, it is reasonable to hypothesize that it must be mediated by a well-established biological process." Next, they declared their intent: "In this report we examine the relationship between the serotonin (5-HT) transporter, the state of being in love and obsessive-compulsive processes."

The preliminaries being out of the way, they got down to business.

Before we get down to business, here is a quick word about technical matters: the chemical they mention – serotonin (5-HT), is involved in regulating all sorts of behavior, including appetite, sleep, arousal, and depression. It is the same chemical that caught the fancy of Professor Peter Fong of Gettysburg University, in Gettysburg, Pennsylvania, whose experiments with feeding Prozac to clams earned him an Ig Nobel Prize in 1998. (See the section of this book titled "The Happiness of Clams.")

Doctors Marazziti, Rossi, Cassano, and Akiskal simplified the entire romantic/obsessive/compulsive morass down to two simple questions:

(1) Is romance literally in people's blood? And, if so,

(2) Is it similar to what's in the blood of obsessive-compulsives?

They already knew that the two-headed monster of obsession and compulsion does, in a measurable sense, flow through the bloodstream. Other scientists had shown that people with obsessive-compulsive disorder have a very different amount of serotonin in their blood than do their non-obsessive, non-compulsive neighbors.

The investigation, then, would be straightforward. They would look at people who were suffering from obsessive-compulsive disorder, and also look at people who were suffering the transports of romantic love. They would compare the blood of both groups with the cooler, laid-back, more prosaic blood of steady, run-of-the-mill, not-in-love, non-obsessive, non-compulsive Janes and Joes.

They decided to examine 20 of each kind. It was easy to find 20 obsessive-compulsives, and also 20 dull people. The task of finding 20 people in love, though, was tricky, because there was no established scientific definition of "romantic love."

Of necessity, Doctors Marazziti, Rossi, Cassano, and Akiskal devised their own definition. This is how it appeared in their published report:

"20 subjects (17 female and three male, mean age: 24) who had recently fallen in love, were recruited from medical students, by means of advertisement. They were selected according to the following criteria:

(a) the love relationship had begun within the previous six months;

(b) the couple had had no sexual intercourse; and

(c) at least four hours a day were spent thinking of the partner."

This definition later proved controversial (see below).

The blood tests gave results that Doctors Marazziti, Rossi, Cassano, and Akiskal found stunningly clear:

"The statistically significant decrease in the [blood levels] of subjects who were in love and in those of obsessive-compulsive disorder patients would seem to suggest a certain similarity between the two conditions ... It would suggest that being in love literally induces a state which is not normal – as is indeed suggested by a variety of colloquial expressions used throughout the ages in different countries, all of which refer generally to falling 'insanely' in love or to being 'lovesick.'"

Dr Marazziti and her colleagues also looked at what happened after the first blush of romance had dimmed. A year after the first blood tests, they interviewed the lovebirds and took new blood samples. Six were still in love with the same people, but no longer thought about their partners day and night. The blood of these six people had become similar to the dull blood of old married couples. Once again, science seemed to confirm what the poets of antiquity knew so well.

For exploring the chemistry of romance and the romance of chemistry, Donatella Marazziti, Alessandra Rossi, Giovanni B. Cassano and Hagop S. Akiskal won the 2000 Ig Nobel Prize in the field of Chemistry.

Donatella Marazziti planned to attend the Ig Nobel Prize Ceremony at her own expense. However, her husband fell ill, and Dr Marazziti sent her acceptance speech via audio tape. In it, she said:

IG NOBEL PRIZES

IS SEX NECESSARY?

Doctors Marazziti, Rossi, Cassano, and Akiskal were well aware that some people consider sex to be part of romantic love. They themselves might not disagree on a personal level, but for scientific purposes, they decided to consider only "subjects who had recently fallen in love and were still at the early, romantic phase of the relationship with no sexual intercourse."

Their report explains why:

"Some might consider sexual intercourse as a necessary component of love. We think not. Stendhal, the French writer, considered love as unconsummated passion. This aspect, we believe, underlies the obsessive preoccupation so characteristic of the early stage of love (which, in rare instances, might persist for a lifetime of abstract idealization that leads to poetry and music dedicated to the love object)."

"Research on love is very important, because love is the engine of human life and of the universe. However, I'm sure that despite all of our efforts, the secrets of nature will remain elusive. I only provide this small insight into the biological mechanism of this typically human feeling. The main bias of my research was that the sample was constituted mainly of Italians, and the Italian way of falling in love may be quite different than that of other populations, such as the Americans. I regret not to be with you, and I send you my best regards and wishes for a joyful ceremony. Please continue to enjoy life, and continue to fall in love."

Marriage By the Million

"MASS PRODUCTION. The term mass production is used to describe the modern method by which great quantities of a single standardized commodity are manufactured. As commonly employed it is made to refer to the quantity produced, but its primary reference is to the method. Mass production is not merely quantity production, for this may be had with none of the requisites of mass production ... Mass production is the focusing upon a manufacturing project of the principles of power, accuracy, economy, system, continuity, and speed."

–from the *Encyclopedia Britannica*, 1926

THE OFFICIAL CITATION

THE IG NOBEL ECONOMICS PRIZE WAS AWARDED TO

The Reverend Sun Myung Moon, for bringing efficiency and steady growth to the mass-marriage industry, with, according to his reports, a 36-couple wedding in 1960, a 430-couple wedding in 1968, an 1,800-couple wedding in 1975, a 6,000-couple wedding in 1982, a 30,000-couple wedding in 1992, a 360,000-couple wedding in 1995, and a 36,000,000-couple wedding in 1997.

Reverend Moon's web site (www.unification.net) is a good, central source of information about mass marriages.

In the early decades of the 20th century, Henry Ford brought mass production to the automobile industry, transforming a small sector of the economy into a model of efficiency and steady growth. In the late decades of the century, another man followed in Ford's footsteps, bringing power, accuracy, economy, system, continuity, and speed to another industry in which it had been lacking.

Reverend Sun Myung Moon held his first mass wedding in 1960, joining 36 couples in a single go. Though small by the standards of later years, this event exemplified the basic business principle that others so often preach and so seldom practice: "Love Your Customer". Reverend Moon never tired, afterwards, of telling audiences: "I never paid a lot of attention to my own children; of course I love them, but I paid more attention to the 36 couples and gave them precedence."

Word of mouth about this standard of lavish customer care spread far and wide. By 1968, the wedding size had increased to 430 couples – an impressive tenfold growth in less than a decade of operation. Of this rapid success, Reverend Moon said that, "the way opened for every person in the world to connect to" his organization.

Henry Ford had based his system on the use of interchangeable parts. Reverend Moon was aware of Ford's success, and as his operational volume increased he adopted Ford's principle. To make it easy for anyone to buy into a mass marriage, Reverend Moon offered to supply a spouse on short notice for whoever wanted one – and no matter what it cost, he consistently met that promise. He perfected the "just-in-time" manufacturing principle that other industries would claim to "discover" some 20 years later.

The business expanded steadily, passing the 1,800-couple-wedding mark in 1975. On that happy occasion Reverend Moon introduced a new slogan to help his new customers adjust to their new possessions:

"Korean saying: 'Ideal mate will come out from a strange place.' ['Strange' and 'ideal' have the same pronunciation in Korean.] So stretch your arms out wide as you can; accept any kind of person." His word was their command, and people accepted whatever spouse Reverend Moon supplied, regardless of what they looked like or what language they spoke.

Business kept on growing. 1982's event featured 6,000-

couples. Reverend Moon introduced a new level of marketing excitement by making the first public announcement about it just 20 days before the wedding occurred. This created a degree of frenzy that continued long after the wedding day. Reverend Moon enjoyed telling crowds: "Those who come to see the mass wedding are surprised, saying, 'I thought the mass wedding was not a big deal – but wow! It is true.'"

The 30,000-couple event ten years later was the result of hard work and careful planning, and demonstrated how a shrewd investment in technology can bring a big payoff. The wedding was televised around the world, using the latest satellite broadcast equipment. Market awareness increased to an almost fantastic degree.

Now the organization was ready to seriously ramp up its scale of production. They did it carefully and well. The result: a flawless 360,000-couple wedding in 1995. Reverend Moon unflaggingly worked at spreading the excitement to people in potential new market segments. At public appearances, he would say:

"Why do we perform the mass wedding? If I explained everything, it would melt the bones of even the grandmothers. They would pine, 'If I could be young again, I would go to the Unification Church mass wedding.'"

Public demand kept growing. On the production side, economies of scale made it possible to rapidly increase output by another factor of ten. A successful 36,000,000-couple wedding in 1997 impressed analysts, and left the competition wondering whether on earth they could compete.

For bringing logarithmic growth to a stagnant industry, Reverend Sun Myung Moon won the 2000 Ig Nobel Prize in the field of Economics.

The winner could not, or would not, attend the Ig Nobel Prize Ceremony.

Thereafter, the operation continued to multiply its output. On February 16, 2002, Reverend Moon and his staff rolled out an all-new, 400-million-couple wedding. With it, they achieved a market penetration just shy of 10%, leaving all competitors effectively in the dust.

Reproduction – Techniques

Clever research and development work has given some new meanings to the time-worn phrase "The Miracle Of Birth." Here are two arguably miraculous achievements:

- High Velocity Birth
- Taking Fatherhood in Hand

High Velocity Birth

"The present invention relates to apparatus which utilizes centrifugal force to facilitate the birth of a child."

–from US Patent #3,216,423

THE OFFICIAL CITATION

THE IG NOBEL MANAGED HEALTH CARE PRIZE WAS AWARDED TO

The late George and Charlotte Blonsky of New York City and San Jose, California, for inventing a device (US Patent #3,216,423) to aid women in giving birth; the woman is strapped onto a circular table, and the table is then rotated at high speed.

Childbirth can be slow and distressing. Inspired by elephants, a childless New York City couple designed a massive electromechanical device that considerably speeds up the process.

George Blonsky was a trained engineer – a mining engineer, as it happened – with an unusually cultivated taste for adventure and invention. Before moving to New York City, he and his wife Charlotte had owned and operated gold and tungsten mines in several parts of the world. George was forever inventing things, though not all of his creations made it past the blueprint stage. George and Charlotte loved children, though they had none, and had written several children's books, none of which were ever published.

They also loved the Bronx Zoo. One day George happened upon the sight of a pregnant elephant slowly twirling herself in circles, evidently in preparation for delivering a 250-pound baby.

The anatomical physics of it galvanized George Blonsky. George performed a simple technical analysis, and discerned the basic principles at work. He then wondered, as engineers will, whether his new-gained technological insight could somehow benefit humanity. Yes, it could, he decided. Yes, it could.

Thus was born the idea of the Blonsky device.

In their patent application, George and Charlotte explained the need for their invention:

"In the case of a woman who has a fully-developed muscular system and has had ample physical exertion all through the pregnancy, as is common with all more primitive peoples, nature provides all the necessary equipment and power to have a normal and quick delivery. This is not the case, however, with more civilized women who often do not have the opportunity to develop the muscles needed in confinement."

And, therefore, George and Charlotte wrote, they would provide "an apparatus which will assist the under-equipped woman by creating a gentle, evenly distributed, properly directed, precision-controlled force, that acts in unison with and supplements her own efforts."

The heart of the idea they expressed in a mere nine words: "The fetus needs the application of considerable propelling force."

George and Charlotte knew how to supply that propelling force.

The rest of their patent – eight very detailed pages altogether – specified exactly how to do it. The design includes some 125 basic components, including bolts, brakes, wing nuts, a massive concrete floor slab, a vari-speed vertical gear motor, a speed reducer, more wing nuts, sheaves, stretchers, shafts, thigh members, a butt plate, aluminum ballast water boxes, still more wing nuts, pillow clamps, a girdle member, and some additional wing nuts.

The patent specifies, in words and diagrams, how the parts are to be combined. Each is numbered for clarity. For example:

"The body of the mother is firmly held in position against movement as a whole under such forces by the boot members (73), the thigh holders (68), the girdle (61), the hand grips (79), and the belts (82), (83) and (84)."

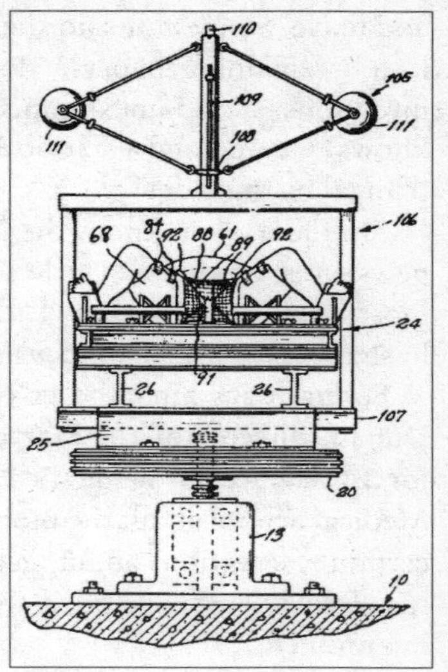

A side view of the Blonsky device. Note the small net for catching the child as it emerges. Some engineers feel it may be inadequate to the task, thus constituting a weak point in an otherwise admirable design.

The Blonskys sent the completed patent application off to Washington. On November 9, 1965, the United States Patent Office granted them the patent for what would henceforth be officially called an "Apparatus for Facilitating the Birth of a Child by Centrifugal Force."

For conceiving what appears to be the greatest labor-saving device ever invented, George and Charlotte Blonsky won the 1999 Ig Nobel Prize in the field of Managed Health Care.

George Blonsky died in 1985, and Charlotte passed away in 1998, just a year before the Ig Nobel Board of Governors honored their achievement.

The Blonskys' niece, Gale Sturtevant, flew 3,000 miles from northern California to the ceremony at Harvard, at her own expense, to accept the Prize in honor of her aunt and uncle. Sturtevant said she has all of George and Charlotte's remaining papers, along with models of various inventions, stored unexamined in her garage. So far as she knows, George and Charlotte never built a full-scale centrifugal birthing device.

"In theory, you know, the idea might work," she told a newspaper reporter. "Uncle George was undoubtedly the most intelligent person I have ever met," added her husband, Don. "His mind was always active."

Several days after the Ig Nobel Ceremony, Dr Andrea Dunaif, director of the Harvard Medical School's Center for Excellence in Women's Health gave a lecture at the Medical School about the Blonsky device. While expressing certain reservations about technical aspects of the apparatus, Dr Dunaif concluded that the Blonskys "were well intentioned."

In subsequent months, the Ig Nobel Board of Governors heard from several women in the final stages of pregnancy. All delivered much the same message. "I know most people think that machine is funny, and so do I," said one, "but after nine months I'm really bored and tired of waiting for this birth. If that machine were available, I'd use it."

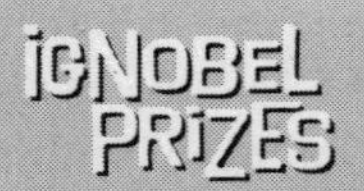

SAFETY FIRST

The Blonskys carefully designed their device to ensure the safety of both mother and child. The machinery includes something known as a "speed governor," which ensures that neither mother nor child can be subjected to a dangerously strong level of force. When operating at its maximum spin rate, the machine would produce a force of seven g – seven times the normal force of gravity. (Note: pilots of jet fighter aircraft typically black out at a force of around five g.)

Taking Fatherhood in Hand

"Anyone who got his sperm is lucky."

–Cecil Jacobson's wife, Joyce, in an interview with *People* magazine

THE OFFICIAL CITATION

THE IG NOBEL BIOLOGY PRIZE WAS AWARDED TO

Dr Cecil Jacobson, relentlessly generous sperm donor, and prolific patriarch of sperm banking, for devising a simple, single-handed method of quality control.

The book *The Babymaker: Fertility Fraud and the Fall of Dr Cecil Jacobson*, by Rick Nelson, Bantam Books, 1994, tells much of the story of Dr Jacobson and his adventures.

Dr Cecil Jacobson was a family man in oh, so many ways, some pleasing, some perhaps otherwise. On the one hand, there were the babies that never existed – Dr Jacobson told hundreds of women they were pregnant when, in fact, they were not. On the other hand, although Dr Jacobson never mentioned it to his patients who did have babies, he was the biological father of a remarkable number of their children. Dr Jacobson's own wife claimed to be pleased and proud about this.

Dr Cecil Jacobson ran a fertility clinic in Vienna, Virginia, a prosperous suburb of Washington, DC. He specialized in helping women get pregnant.

The clinic consisted of Dr Jacobson as the only physician, and a varying staff of drudge workers including his wife Joyce, some of their own large brood of children and, until the money ran short, a very small number of other administrative workers.

Customers flocked to the clinic. Dr Jacobson had a considerable reputation, based on some very real early accomplishments followed by years of loud, quasi-accurate boasting. When a young man, Dr Jacobson was among the pioneers in using amniocentesis to diagnose the health of a developing fetus. After that, though, he moved from job to job, from hospital to hospital, as one boss after another decided that the celebrated Dr Jacobson was ever more a creature of words than of genuine accomplishments. After being quietly but forcefully terminated from these places of employ, he would falsely claim to be still affiliated with them.

Couples who have difficulty conceiving a child sometimes grow desperate. In the Washington area, Dr Jacobson was the one to whom the most anxious, the apparently hopeless cases would be referred.

Unlike other doctors who deal with pregnancy and childbirth, Dr Jacobson seldom gave his new patients any kind of examination. Nor, typically, did he ask many questions. What he did was talk.

Dr Jacobson was a real talker. He'd tell a women he was going to get her pregnant. Sometimes he would guarantee it. He'd instruct her to have sex with her husband diligently and often. He insisted that some couples have sex every day for weeks on end.

The genius lay in what else he did. He would inject the woman with a generally innocuous hormone called Human Chorionic Gonadotrophin (HCG). This accomplished two good things for Dr Jacobson. First, it brought in lots of money, because he charged for every injection and he would give patients dozens and dozens of them. Dr Jacobson bought so much HCG that his tiny one-doctor clinic was reportedly one of the largest purchasers of HCG in the world; he bought at very low cost and then marked up the price tremendously.

The HCG injections also brought delight, albeit temporarily, to the patients. All that HCG in their bloodstreams caused the women to register positive on a simple pregnancy test. In at least one case, Dr Jacobson told a 49-year-old post-menopausal woman that she would get pregnant, then he injected her with lots of HCG, and voila, a pregnancy test indicated she was with child.

After announcing the "good news," Dr Jacobson would use ultrasonography to produce a fuzzy – extremely fuzzy – "picture" of what he said was a fetus. This sent the supposedly imminent parents into giddy rapture. He would then take new sonograms over the succeeding weeks and months, each time producing an indecipherably fuzzed sonogram image that supposedly showed a developing fetus.

Dr Jacobson would warn the supposedly pregnant women not to go see their regular obstetricians, saying it would somehow endanger the pregnancy. Patients who, despite the warning, did go see their own doctors got a jarring surprise: they were not pregnant. When a distressed women would then ask Dr Jacobson how this could be, he explained that the pregnancy had spontaneously aborted, and that the fetus had been completely "absorbed" into the mother's body, leaving no trace. He would then recommend a new round of HCG injections and regular sexual intercourse. Many of his patients followed that advice, repeatedly, enduring years of heartbreak.

All that would have been enough to earn Cecil Jacobson a place in history. It was enough to get the press and the police interested. But Dr Jacobson's greatest fame came from an additional little service he provided to some of his patients.

Some woman did conceive while under Dr Jacobson's care. These women he did immediately send back to their own obstetricians to oversee the pregnancy. Some of these

The Babymaker was written by Rick Nelson, one of the journalists who first uncovered the real story of Dr Cecil Jacobson. Nelson himself played a small, curious role in the story. After learning that Dr Jacobson was injecting women with hormones that would cause them to falsely test positive on pregnancy tests, Nelson had himself injected with the same hormone. He then took a pregnancy test. The test result indicated that Mr Nelson was pregnant.

pregnancies were the result of nature taking its course, others started with artificial insemination that had been recommended and performed by Cecil Jacobson.

Dr Jacobson told each of the artificial insemination patients that he used anonymous sperm donors, whose physical characteristics he matched to those of the woman's husband. In fact, Dr Jacobson himself donated all of the sperm.

In 1991, Cecil Jacobson was brought up on 53 counts of fraud. Prosecutors made it clear that they were presenting just the tip of the iceberg. They estimated that Dr Jacobson's sperm, hand delivered by Dr Jacobson, had produced as many as 75 children.

To the end, a small band of supporters, including US Senator Orrin Hatch from Cecil Jacobson's native state of Utah, maintained that the doctor was a saintly, heroic man who had been unfairly maligned and persecuted.

Dr Jacobson was convicted on all counts, and got a prison sentence. He also got the 1992 Ig Nobel Prize in the field of Biology.

The winner could not, or would not, attend the Ig Nobel Prize Ceremony. He had a previous five-year engagement.

Reproduction – Equipment

To do any job well, one must have the proper equipment, understand how it works, and keep it in good repair. This chapter chronicles two projects in which the equipment was the focus of attention:

- Insert Here

- Height, Penile Length, and Foot Size

Insert Here

"OBJECTIVE: To find out whether taking images of the male and female genitals during coitus is feasible and to find out whether former and current ideas about the anatomy during sexual intercourse and during female sexual arousal are based on assumptions or on facts.

METHODS: Magnetic resonance imaging was used to study the female sexual response and the male and female genitals during coitus. Thirteen experiments were performed with eight couples and three single women.

CONCLUSION: Taking magnetic resonance images of the male and female genitals during coitus is feasible and contributes to understanding of anatomy."

–from the report by Schultz, van Andel, Mooyaart, and Sabelis

THE OFFICIAL CITATION

THE IG NOBEL MEDICINE PRIZE WAS AWARDED TO

Willibrord Weijmar Schultz, Pek van Andel, and Eduard Mooyaart of Groningen, The Netherlands, and Ida Sabelis of Amsterdam, for their illuminating report, "Magnetic Resonance Imaging of Male and Female Genitals During Coitus and Female Sexual Arousal."

Their study was published in the *British Medical Journal*, vol. 319, 1999, pp. 1596–1600. The full version of Ida Sabelis's first-hand account of her experience was published in the *Annals of Improbable Research*, vol. 7, no. 1, January/February, 2001, pp. 13–14.

A research team in The Netherlands gave humanity its first good, inside look at a couple's genitals while those genitals were in use.

It is possible to fit two people inside the cylinder of a Magnetic Resonance Imaging machine, and this in turn makes it possible to take MRI images of the couple's sex organs in operation, if the two people are not claustrophobic.

The research team recruited several couples who were willing and able to perform under these technologically cloistered conditions, and who had all the requisite equipment including, in the case of the females, an intact uterus and ovaries. The scientists assured the participants that there would be confidentiality, privacy, and anonymity. (After the couples had donned their clothing, though, some of them happily chose to shed their anonymity.)

The set-up had a distinctly clinical feel: "The tube in which the couple would have intercourse stood in a room next to a control room where the researchers were sitting behind the scanning console and screen. An improvised curtain covered the window between the two rooms, so the intercom was the only means of communication."

The arrangement was in some ways not unlike those of early NASA astronauts in their space capsules, exchanging instructions and acknowledgements by radio link. In other ways, the arrangement was unlike those of early NASA astronauts in their space capsules.

The experimental procedure was straightforward:

"The first image was taken with [the woman] lying on her back. Then the male was asked to climb into the tube and begin face to face coitus in the superior position. After this shot – successful or not – the man was asked to leave the tube and the woman was asked to stimulate her clitoris manually and to inform the researchers by intercom when she had reached the preorgasmic stage. Then she stopped the autostimulation for a third image. After that image was taken the woman restarted the stimulation to achieve an orgasm. 20 minutes after the orgasm, the fourth image was taken."

That's all there was to it.

Six of the couples succeeded in at least partial penetration. Two couples were invited to repeat the procedure one hour after the man had taken Viagra. Both couples accepted the offer, with positive results.

The overall results were impressive. As the researchers described them in their published report:

"The images obtained showed that during intercourse in the 'missionary position' the penis has the shape of a boomerang and ⅓ of its length consists of the root of the penis. During female sexual arousal without intercourse, the uterus was raised and the anterior vaginal wall lengthened. The size of the uterus did not increase during sexual arousal."

The scientists were quietly triumphant, because their objectives were largely met. They *did* find out whether taking images of the male and female genitals during coitus is feasible, and they *did*, to some limited extent, find out whether former and current ideas about the anatomy during sexual intercourse and during female sexual arousal are based on assumptions or on facts.

For their contributions to the study of anatomy and physiology, Willibrord Weijmar Schultz, Pek van Andel, Eduard Mooyaart and Ida Sabelis won the 2000 Ig Nobel Prize in the field of Medicine.

Pek van Andel traveled, at his own expense, from Groningen, The Netherlands, to the Ig Nobel Prize Ceremony. In accepting the Prize, he said:

"To find something truly new you need an unpredictable element: a strange observation, idea or experiment. The new comes – by definition – by surprise. When I saw an odd scan of the throat of a singer, singing an 'Aaa...,' I wondered: Why not a scan of a love act?

"The hardware was no problem. Removing the table from the scanner tube gave enough space to make love. The software? No problem, we programmed the machine to do a scan of a 'patient' of '300 pounds.' The wetware? We had enough slim volunteers. The only problem was the Red Tape, so we had to do our work clandestinely.

"Our first scans revealed an amazing anatomy. A penis as a boomerang with a huge root. An unchanged uterus and a fast-filling urine bladder!

"Three times our article was refused for publication. Twice by *Nature* magazine and once by the *British Medical Journal*. Finally the *British Medical Journal* took it, after a check in Holland, behind our back, as to whether the article was fake or not. The lesson of this is you should cherish your idiot idea, and push it through – even through your boss or bosses, if need be."

The day after the Ig Nobel Ceremony, Dr van Andel lectured at Harvard Medical School, giving technical information and advice for those who would like to undertake this line of research. The audience of doctors was hushed and amazed.

To Make Love As a Testee

Ida Sabelis is an organizational anthropologist at Heemstede, The Netherlands. On the occasion of winning the Ig Nobel Prize, she prepared an account of her role in the experiment. This is a shortened version. It was translated from Dutch into charming English by Truus Pinkster. (Note: This account works especially well when read aloud to an audience, for educational or other purposes.)

"In the autumn of 1991, Pek (M.V. van Andel) phoned my partner Jupp. Whenever he does that, he mostly has something special on his mind. The point was to visualize with a modern scan how it really shows when a man and a woman are making love. It would be very beautiful to image this act.

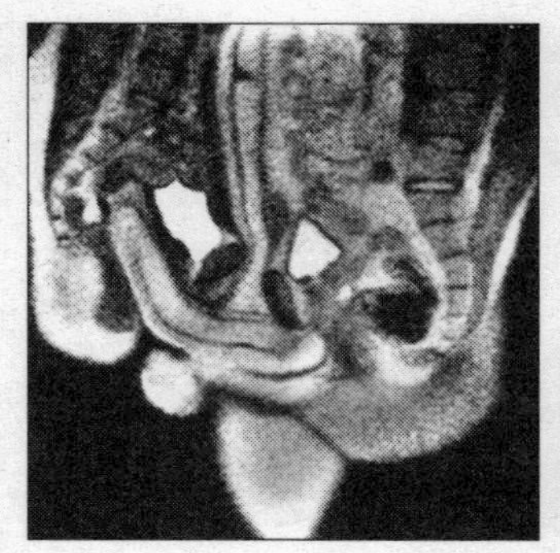

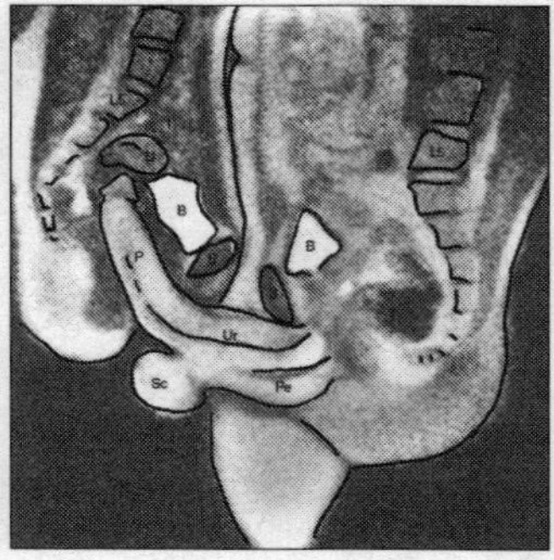

An original image published in the *British Medical Journal* from the MRI session, and a copy in which the investigators have helpfully outlined the areas of special interest.

Pek suggested it should be just something for us, we are slim, and because of our background as acrobats.

"My background in women's movement doesn't give me much reason to trust in advance in the humane feeling of medical men, especially medical specialists. And, as everybody knows, shots of love acts can be used for goals for which I wouldn't be available. But after the first conversations with Weijmar Schultz and the other 'medical gentlemen' very quick there is a good atmosphere.

"From the control room you have a sight through a window on the big white space where the MRI machine is arranged. In the centre of the enormous cake tin is a tube in which people can be slide in and out by a sledge-bed. The tube is circa 60 centimetres and the height on the highest point nearly 35 centimetres. Embarred through the spectacle I withhold my thought that we probably never can embark that machine. We agree that anyhow we will try.

"But the most and most important thoughts preceding the happening yet were on ourself: how shall it be in such a sterile white tube? Would we able to switch off the surrounding and just have a good time? What shall we do when one of us shall get not any sexual arousal in 'that thing?' How can we help each other to do that for which we have come? Should we nearly be stucked in that tube or still have any 'play-room?'

"We undress ourselves, lying down on the sledge-bed and are slided in by Eduard: we are lying on our side and facing each other ... That's the position unto which we decided and which comes the nearest to the expectation: of the photographers gentry: on top of each other, male on top, female under. In every way we reject the idea to lie on top of each other and packed together, much too heavy and moreover a position which for me produces hardly any arousal.

"It's narrow in the tube – there's nothing other to expectate – but it goes ... I can manoevre my left hand just to the place where I wish him and above us the pitching of the magnets starts.

"Then nothing for a little while. Confined by the space we make the best of it and that's just not so uncosy. On a certain moment there's sounding through the microphone 'the erection is fully visible, including the root.' Again nothing for a little while. We report the control room that their microphone has to be open as we don't know what's going on. The first shots are taken: 'Now lay down very still and holding your breath during the shot!' Forty rhythmic little bangs of the magnets above us and then one may exhale.

"We are giggling a lot, because excitement, let alone of which nature, and an erection as the most visible, simply sinks down like an arrow when you have to hold your breath during many seconds ... and then going on.

"It's becoming pleasantly warm in the tube and we truly succeed in enjoying each other from time to time in a familiar way. When the microphone is telling us that we may come – insofar possible – and that only we have to inform them in connection to the photo, we burst out into a roar of laughter and some moments later we do what is the purpose. Sniggering we lay down a while before we announce that we just now like to go out. Like buns which are pushed out from the oven, we are coming outside.

"Enthusiasm everywhere, it works and we get dressed quickly to look at the shots in the control room. Of course, some are blurred because of movement. But some other are of an amazing beauty: that we are! Not so much a passport photo for daily use but surely a shot that shows so much that it makes me speechless.

"There, it's my womb and surely, on that place is Jupp, naturally in a way as I know from my own sensation: below the cervix. Very clear all details of our both innersides are visible up to and including the common boundary between our both bellies.

"Only two days later I'm feeling a kind of pride: we tried and succeeded!"

Height, Penile Length, and Foot Size

"To determine whether 'folk myths' regarding the relationships of penile size to body height and foot size have any basis in fact, 63 normally virilized men were studied. Height and stretched penile length were measured; shoe size was recorded and converted to foot length. Penile length was found to be statistically related to both body size and foot length, but with weak correlation coefficients. Height and foot size would not serve as practical estimators of penis length."

–from Bain and Siminoski's report

THE OFFICIAL CITATION

THE IG NOBEL STATISTICS PRIZE WAS AWARDED TO

Jerald Bain of Mt Sinai Hospital in Toronto and Kerry Siminoski of the University of Alberta for their carefully measured report, "The Relationship Among Height, Penile Length, and Foot Size."

Their study was published in *Annals of Sex Research*, vol. 6, no. 3, 1993, pp. 231–5.

Science at its best determines whether what "everyone" believes is in fact true. Dr Jerald Bain and Dr Kerry Siminoski examined one of mankind's cherished and feared beliefs. They attacked this problem with a ruler.

"One of the more prevalent beliefs," wrote Dr Bain and Dr Siminoski, "involves the theory that the size of a man's penis may be estimated indirectly by assessing overall body size, or by gauging the size of another of his appendages (such as his ear lobes, nose, thumbs, or feet), and extrapolating to penile length. Depending on the underlying hypothesis, the penis is assumed to correlate either directly or inversely with the dimensions of one of these other body parts. To scientifically address this, we studied the relationships among penis length and two of

Dr Jerald Bain accepts his Ig Nobel Prize. Photo: Relena Erskine/ Anna Boysen/*Annals of Improbable Research*.

these anatomic variables, overall body height and foot length."

To do this, Dr Bain and Dr Siminosky recruited 63 men who were willing to have their pertinent body parts measured. In their report, Dr Bain and Dr Siminoski do not specify what method they used to recruit the men.

Dr Bain and Dr Siminosky measured the body parts. The heights ranged from 157 to 194 centimeters. The feet ranged from 24.4 to 29.4 centimeters. The penises ranged from 6.0 to 13.5 centimeters. Penis length was measured while the penises were stretched. In their report, Dr Bain and Dr Siminoski do not specify what method they used to stretch the penises.

With the data in hand, Dr Bain and Dr Siminosky did a statistical analysis. Dr Bain and Dr Siminoski specify that the method they used was a least-squares linear regression.

Their analysis indicated that there is what they call a "weak" correlation between a man's height and the length of his penis, and that there is also a "weak" correlation between his foot size and his penis length.

Their final conclusion: "Our data ... indicate that there is no practical utility in predicting penis size from foot size or height."

For making statistics interesting to the common man, Dr Jerald Bain and Dr Kerry Siminoski won the 1998 Ig Nobel Prize in the field of Statistics.

Dr Bain traveled, at his own expense, from Toronto to the Ig Nobel Prize Ceremony. In accepting the Prize, he said:

"This is a real study, and it's a very important study, and

As Harvard physics professor Roy Glauber sweeps paper airplanes from the stage, Nobel Laureates (right to left) Richard Roberts, William Lipscomb, and Dudley Herschbach display their gigantic footwear. Sheldon Glashow (visible over Glauber's shoulder) rushes to join them. The Laureates took their stand as a tribute to Bain and Siminoski's Prize-winning report, "The Relationship Among Height, Penile Length, and Foot Size." Photo: Eric Workman/*Annals of Improbable Research*.

I hope you'll all take it seriously. There has been an old folk mythology about the relationship between certain bodily appendages and foot size. Now, I wasn't really aware of this mythology until some years ago when my late mother-in-law – my dear late mother-in-law – she was a wonderful woman – I loved her dearly, I think she loved me dearly…

"One day after we had three children – we now have four – my mother-in-law said to my wife, 'Did you see how small Jerry's feet are?' So Sheila, my dear wife, said, 'So?' And her response was, 'Don't you know?'

"This research was spurred on to answer the eternal question. I must tell you that the answer is that, yes, there is a very – and it's hard to find the right words when talking about this – there is a very weak relationship. But there is also a relationship between penile size and height, so that taller, bigger people – it's just a reflection of body size – but, take heart, take heart, because it would be foolish and frivolous of any woman to try to judge the potential size of the – it would be very difficult, because the most important fact to know is that the erection is the great equalizer."

The following day, Dr Bain lectured at Harvard, explaining his fascination and expertise in penile and other body-part measurement, and illustrating his points with colorful slides, statistics, and personal memorabilia.

Discoveries – Basic Science

Scientific breakthroughs are called "breakthroughs" because they are so surprising. They come against and break though a great wall of expectation. This chapter describes six very surprising discoveries that were eventually, and inevitably, honored by Ig Nobel Prizes:

- A Bug in His Ear
- The Happiness of Clams
- To Boldly See What Others Don't
- Cold Fusion in Chickens
- Mini-Dinosaurs, Mini-Princesses
- The Remembrance of Water Passed

A Bug in His Ear

"The sounds in my ear (fortunately I had chosen only one ear), were becoming louder as the mites traveled deeper toward my ear drum."

–from the letter "Of Mites and Man"

THE OFFICIAL CITATION

THE IG NOBEL ENTOMOLOGY PRIZE WAS AWARDED TO

Robert A. Lopez of Westport, New York, valiant veterinarian and friend of all creatures great and small, for his series of experiments in obtaining ear mites from cats, inserting them into his own ear, and carefully observing and analyzing the results.

His letter, entitled "Of Mites and Man," was published in *Journal of the American Veterinary Medical Association*, vol. 203, no. 5, Sept. 1, 1993, pp. 606–7.

Some of the greatest medical advances are made by brave doctors who experiment on themselves. One of the bravest, boldest, and most memorable medical experiments was conducted by Dr Robert A. Lopez in 1968. Here is the description he wrote some years later, when he had had sufficient time to reflect on the experience.

Of Mites and Man
by Robert A. Lopez, DVM
Westport, New York

"Two strange and related clinical cases prompted me to investigate the possibility of transmission of the ear mite, Otodectes cynotis, to human beings. In the first case, a client, accompanied by her three-year-old daughter, brought in two cats with severe ear mite infestations. In the examining room, the daughter happened to complain of itching chest and abdomen. The mother stated that the daughter frequently held the cats for long periods, like dolls, then showed me the numerous small red abdominal bite marks that were the source of the itching. I recommended that she check with her pediatrician. After the cat's ear mite infestation was cleared up, I learned that the daughter's itching also quickly disappeared. A year later, when the same client brought in a cat heavily infested with ear mites, she complained of bites on her ankles. The bites subsequently stopped after the cat had been cleared of the ear mite infestation.

"At that time (1968), a search of the literature did not reveal any report of Otodectes cynotis infestation in human beings, so I decided to be a human guinea pig.

"I obtained ear mites from a cat and confirmed by microscopic examination that they were Otodectes cynotis. Then I moistened a sterile, cotton-tipped swab with warm tap water and transferred approximately one g of ear mite exudate from the cat to my left ear. Immediately, I heard scratching sounds, then moving sounds, as the mites began to explore my ear canal. Itching sensations then started, and all three sensations merged into a weird cacophony of sound and pain that intensified from that moment, 4 p.m., on and on ... At first, I thought this wouldn't, and couldn't, last very long. However, as the day and evening wore on, I began to worry. The pruritus was increasing. The sounds in my ear (fortunately I had chosen only one ear), were becoming louder as the mites traveled deeper toward my ear drum. I felt helpless. Is this the way a mite-infested animal feels?

"For the next five hours, the mites were very active and then their activity, measured by scratching sounds and degree of pruritus, leveled off. There still was something definitely crawling about deep in my left ear, but the discomfort was bearable.

"After retiring about 11 p.m., the mite activity increased incrementally so that, by midnight, the mites were very busy, biting, scratching, and moving about. By 1 a.m., the sounds were loud. An hour later the pruritus was very intense. After two hours the highest level of itching and scratching was attained. Sleep was impossible. Then, suddenly, the mites seemed to lessen their feeding activities. The noise and pruritus abated, and a brief sleep was possible. Mite activity resumed at 7 a.m., with light noises and slight pruritus. This pattern was repeated – light mite activity during the day, with slight increase in the evening, from approximately 6 to 9 p.m., and then heavy mite activity from midnight to 3 a.m. This night feeding pattern was quite regular and made sleep, no matter how demanding, completely out of the question.

"By the second week, when the late night feeding pattern had become well established, the intensity of the mite activity began to lessen. By the third week, the ear canal was filling up with debris, and hearing from my left ear was gone. By the fourth week, mite activity was 75% reduced and I could feel mites crawling across my face at night. They never did try to enter my right ear, nor did they bite or cause any itching anywhere else on my body. At the end of one month, I could no longer hear or feel any mite activity. The pruritus and internal ear noises were going. However, my ear was completely filled with exudate. I cleansed my ear with warm water swabs and flushings, for the first time. Within one week, my left ear was clear of debris. By the sixth week, there was no pruritus, and hearing was normal. Recovery was surprisingly fast with just warm water irrigations.

"By the eighth week, I decided to try again to see whether the first experiment had been flawed or

Dr Robert Lopez, who experimentally put cat ear mites into his own ear, accepts his Ig Nobel Prize. Photo: Stephen Powell/*Annals of Improbable Research.*

misleading. Accordingly, with my left ear now healed, no debris evident, and hearing normal, I obtained ear mites from another cat and confirmed their identity as Otodectes cynotis, as I had done before. I transferred a 1- to 2-g sample of the cat's ear exudate to my left ear, as I had done before. Once again, my ear began to react to the mite invasion. Loud scratching noises, pruritus, and pain all began within a few seconds. The same pattern evolved, with an early evening feeding pattern and a late heavy-eating session. Intermittent feeding forays during the day were short. The first week was again filled with intense pruritus, and the second week with lessened mite activity that ceased by day 14. The left ear was filled with much less exudate, and my hearing was only slightly impaired. Warm water irrigations cleared up the infestation in 72 hours.

"This definite reduction in symptoms left many questions. Was there an immunity? Were human ears refractory to Otodectes? A third and final trial had to be done.

"At week 11, I repeated the experiment as before, using my left ear. Within a few minutes, the pruritus and inner ear noises began. However, this time the intensity was much less severe. Very little debris accumulated, and hearing was only partially affected. Feeding patterns remained the same. By the end of day eight or nine, the mites had ceased biting, although I had felt some walking across my face during the night. Once again, nothing but warm water was used to rinse the left ear. It healed again quickly, except for occasional slight bouts of pruritus.

"This descending time and intensity of infestation or increasing immunity under similar modes of experimentation raises some interesting questions. Is there an immune reaction to parasitisms, particularly Otodectes cynotis, in mammals? In over 30 years of practice, I have noticed that the younger cats had more severe ear mite infestations.

"Do otodectic mites have a regular feeding pattern? If so, would late evening treatments be more effective? I routinely advise clients to use ear medications for mites late in the evenings.

"Since my initial literature search, I have found one report of natural ear mite infestation in a human being [published in 1991 in a Japanese medical journal], causing tinnitus. I wonder whether the person involved enjoyed her experience as much as I did."

For his services to man, mite, and kittens, Robert Lopez won the 1994 Ig Nobel Prize in the field of Entomology.

Dr Robert Lopez traveled to the Ig Nobel Prize Ceremony, at his own expense. His acceptance speech took the form of a poem:

"I hate the old didactic mite.
All he does is crawl and bite.
At sleeping time he acts just like a bum,
Crawls right into your ear drum.
Once there, he scratches and bites ad infinitum."

After reciting his poem, Dr Lopez took from his pocket a selection of bugs, which he distributed to the audience. The Ig Nobel Board of Governors was unable to determine which species were involved.

The Happiness of Clams

"At human serotonergic brain synapses, selective serotonin re-uptake inhibitors such as fluoxetine ('Prozac'), fluvoxamine ('Luvox'), and paroxetine ('Paxil') block re-uptake transporters and effectively increase serotonin neurotransmission. In humans, serotonin (5-hydroxy-tryptamine, 5-HT) regulates behaviors such as appetite, sleep, arousal, and depression. In bivalve molluscs, 5-HT has salient effects on reproductive processes such as spawning, oocyte maturation and germinal vesicle breakdown, sperm reactivation, and parturition."

–from the research report by Peter Fong, et al.

THE OFFICIAL CITATION

THE IG NOBEL BIOLOGY PRIZE WAS AWARDED TO

Peter Fong of Gettysburg College, Gettysburg, Pennsylvania, for contributing to the happiness of clams by giving them Prozac.

The study was published as "Induction and Potentiation of Parturition in Fingernail Clams (Sphaerium striatinum) by Selective Serotonin Re-Uptake Inhibitors (SSRIs)," Peter F. Fong, Peter T. Huminski, and Lynette M. D'urso, *Journal of Experimental Zoology*, vol. 280, 1998, pp. 260–4.

Ever since its introduction in 1987, Prozac has been one of the most-prescribed antidepressant drugs. It is used to treat human beings, and sometimes also cats, dogs, and other pets.

Professor Peter Fong gave Prozac to clams, and he had a very good reason for doing so.

It's mostly about sex.

The drug fluoxetine – popularly known as Prozac – helps rouse many patients from a severe depression. That is the main reason for its sudden and great popularity with both doctors and patients. Like many medicines, fluoxetine's effects can be hit-or-miss: for some people it works like magic, for others, it does little or nothing. For some patients the effect is more like hit-and-run: in these people fluoxetine seems to lessen or kill the sexual drive.

But when it comes to sex, there have long been intriguing hints that fluoxetine can work in the opposite direction. A 1993 report in the *Journal of Clinical Psychiatry* concluded that the "Sexual side effects of fluoxetine may be more variable than previously thought." The report's title: "Association of Fluoxetine and Return of Sexual Potency in Three Elderly Men."

So it was perhaps not entirely out of the realm of possibility that when Prozac was fed to clams, it would have some happy effect on the little animals' sex lives. Knowing this possibility, though, did not fully prepare Peter Fong for the magnitude of the effect. When he fed fluoxetine to clams, the results were, in sexual terms, spectacular.

Fong fed the fluoxetine to clams as an experiment. He chose clams because clams and humans (and cows, lobsters, squid, and a multitude of other animals) show some remarkable similarities way deep down in their nervous systems. On the cellular level, much of what transpires in the clam is remarkably like what goes on in people. By studying the nervous system of clams – tinkering, measuring, feeding it Prozac – scientists can sometimes learn surprisingly useful amounts of information about human beings. Experiments on clams often, in addition, can be done more quickly, more cheaply, and with less paperwork than corresponding experiments on humans.

Fong's fluoxetine findings were not without scientific import. Then and now, nobody fully understands how fluoxetine and its chemical cousins work. The actions of the nervous system are complex and subtle, defying anyone who tries to tease out their secrets. Nevertheless, Professor

Peter Fong of Gettysberg College, Gettysburg, Pennsylvania, did find some hidden nuggets.

He found that if you feed Prozac to clams (at least to clams of the species Sphaerium striatinum, also known as "fingernail clams") they begin reproducing furiously – at about ten times their normal rate.

Thus, thanks to Peter Fong, we know that Prozac has measurably profound effects on the nervous system and reproductive behavior of the fingernail clam. For demonstrating this, he won the 1998 Ig Nobel Prize in the field of Biology.

The winner could not attend the Ig Nobel Prize Ceremony, as he had to teach a class that day. Instead, he sent an acceptance speech to be presented at the Ceremony. Here are Professor Fong's words, which were read aloud that evening by Dr Peter Kramer, author of the book *Listening to Prozac*:

"Many people have asked me how I came upon using Prozac to make clams have sex. It happened quite by accident. It was late one night, and I was sitting alone in my laboratory feeling pretty depressed. Rising from my chair, I clumsily knocked over my prescription of Prozac and watched helplessly as several capsules fell into an aquarium full of clams. To my amazement, the clams began spawning copious amounts of sperm and eggs into the water. Suddenly, I was no longer depressed. The rest is history. I thank the manufacturers of Prozac, and I salute the clams – Nobel beasts who gave their lives for my research, but at least they had sex before they died and went out with a bang. Happy as a clam."

Professor Fong continued to do research on other aspects of reproductive biology and the ecology of aquatic invertebrates. In 2001, for example, he published a paper in the

Journal of Experimental Zoology elucidating the mechanism of penile erection in the snail Biomphalaria glabrata.

Professor Fong has not wholly left behind the subject that earned him an Ig Nobel Prize. To the delight of those who savored his work with Prozac and clams, in 2002 he wrote a chapter for the book *Pharmaceuticals and Personal Care Products in the Environment* (published by the American Chemical Society Press) The chapter title: "Antidepressants in Aquatic Organisms."

To Boldly See What Others Don't

"These official mission films – analyzed over a period of four years, via scientific techniques and computer technologies literally unavailable even to NASA 30 years ago (when the original photographs were taken) – now provide compelling scientific evidence for the presence of ancient artificial structures on the Moon. Further, it is now apparent that the entire purpose of President John F. Kennedy's sudden, all-out Apollo Program to land Americans on the Moon within ten years was to send American astronauts directly to these ruins, to record them on film, and to bring back physical evidence (including manufactured artifacts) for analysis on Earth."

–from a press release by Richard Hoagland

THE OFFICIAL CITATION

THE IG NOBEL ASTRONOMY PRIZE WAS AWARDED TO

Richard Hoagland of New Jersey, for identifying artificial features on the moon and on Mars, including a human face on Mars and ten-mile-high buildings on the far side of the moon.

His findings are published in *The Monuments of Mars : A City on the Edge of Forever*, North Atlantic Books, Berkeley, CA, 1987.

Some of the greatest discoveries go overlooked until the arrival of a specially keen observer. Spacecraft sent us thousands of photographs of the moon and Mars. Millions of people gazed in wonder at these highly detailed views of distant worlds. Richard Hoagland gazed, too. Keener than most of his fellow earthlings, he – unlike them – did not fail to recognize the buildings, the giant face, the flying saucer, and all of the other large-scale engineering bric-à-brac.

The structures that Richard Hoagland discovered on Mars, and those he found on the back side of the moon, are too important to be neglected. Because NASA neglects them,

This is the first photograph of the face on Mars. It was taken by the Viking 1 spacecraft in 1976. Photo: NASA.

because most of the news media neglect them, because school curricula neglect them, Hoagland has made it his duty to educate the public. He works at this diligently, tirelessly, as repetitively as necessary, in his best-selling book, in his press releases, in his press conferences, and in his frequent appearance on late, late, late-night radio talk shows.

The most inexplicable thing, to Hoagland, is the face. In 1976, the Viking Orbiter spacecraft sent back wonderful photos of the surface of Mars. In one image, something shaped like a massive face seemed to be staring up at the camera. Word got around, and whispers spread that the face was not a mere haphazard collection of rocks and shadows. To Richard Hoagland, it was as plain as the face under his nose that some civilization had sculpted that face, probably as a tomb or monument, and probably suspecting that one day Richard Hoagland, or someone like him, would come along to see it.

He agreed with skeptics that the camera images were pretty grainy, and suggested that the next generation of spacecraft and cameras would show the face in its true, carefully designed-and-manufactured glorious detail.

In 1987, Hoagland published the book that would gain him fame among those who knew who he was. *The Monuments of Mars: A City on the Edge of Forever* is full of photographs, and chock-a-block with explanations. There is an entire city on Mars, the book explains, with a large pyramid ("a five-sided, bi-symmetrical, 'buttressed', kilometer-high object, located but a few kilometers away from the Face"), a cluster of other pyramids, a fortress, a city square, a "honeycomb" complex located between the fortress and the big pyramid, and a runway complex.

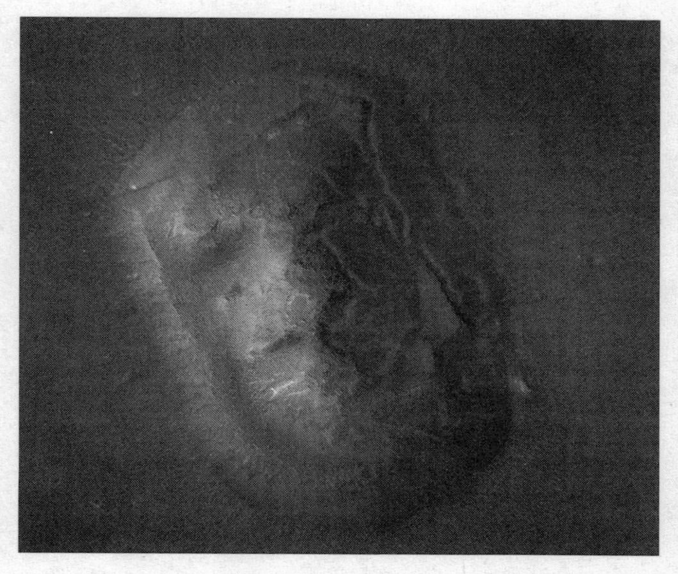

A higher resolution photo taken in 2001. Some wags have suggested that the face got in a terrible fight some time between 1976 and 2001. Photo: NASA/JPL/Malin Space Science Systems.

The runway complex may be used to launch the flying saucer which Hoagland recognized in a different Mars photo, one he saw several years after the book came out.

Richard Hoagland went on to analyze photos from the moon, taken by the Lunar Orbiter III spacecraft. On the back side of the moon he found two objects of more than passing interest. These he described as an approximately "seven-mile-high, glass-like 'Tower/Cube'" and a 1.5-mile-high, glass-like "shard." Both, he explained, are of artificial origin.

In 2001, a new spacecraft began sending back new photos of Mars, of much higher resolution than the grainy pictures of 1976. Just as Hoagland had predicted, the new photos showed the face in much more detail. The rocks and shadows are much more easily seen than in the earlier views. The facial features are much, much subtler, almost to the point of vanishing.

For seeing what others would not or could not, Richard Hoagland won the 1997 Ig Nobel Prize in the field of Astronomy.

The winner could not, or would not, attend the Ig Nobel Prize Ceremony.

He has continued his singular effort to lift humanity from its ignorance of civilizations that are, or were, at work on the moon, on Mars, and perhaps in the minds of men.

Cold Fusion in Chickens

"Thus in the 20 hours that intervened, the hens transformed a supply of potassium into calcium."

–from Louis Kervran's book *Biological Transmutations*

THE OFFICIAL CITATION

THE IG NOBEL PHYSICS PRIZE WAS AWARDED TO

Louis Kervran of France, ardent admirer of alchemy, for his conclusion that the calcium in chickens' eggshells is created by a process of cold fusion.

His study was published as *A la Découverte des Transmutations Biologiques*, Le Courrier des Livres, Paris, 1966. It was translated into English, together with other of Kervran's scientific writings, and published as *Biological Transmutations and their Applications in: Chemistry, Physics, Biology, Ecology, Medicine, Nutrition, Agronomy, Geology*, Swan House, 1972.

Chemistry is not as difficult as students fear. The equations taught in school are wrong. The so-called elements (hydrogen, helium, lithium, beryllium, boron, carbon, nitrogen, and all the rest) are not elemental at all. Transforming one into another – turning silicon into calcium, or manganese into iron – is easy. Happens all the time. Even a chicken can do it. In fact, chickens do do it.

This was Louis Kervran's message.

All of modern chemistry – which is to say all of chemistry since it became a science – is based on the idea that there are stable atoms of different kinds. An atom of iron is different from an atom of chlorine is different from an atom of silver is different from an atom of gold. Chemistry is all about how you stick lots of atoms together in bunches (the technical word is "compounds"), and how you recombine the bunches into other bunches. All the substances we encounter are formed by atoms bunching together this way and that.

All well and good, wrote Louis Kervran, but inside living creatures, "the improbable surely happens." Inside living creatures, atoms aren't limited to just palling around with other kinds of atoms. Inside living creatures, one kind of atom can turn into a different kind. An atom of silicon can become, voila, an atom of calcium. An atom of iron can become an atom of manganese, or vice versa.

For centuries, optimists, cockeyed and otherwise, had hoped and prayed that base elements, such as lead, could be transmuted into expensive ones, such as gold. But no one had ever seen such things happening (except, in a very limited way, in the hellishly super-heated confines of stars and nuclear explosions).

Scientists never noticed it happening, Louis Kervran explained, because they paid too much attention to dead solids and dead liquids and dead gases, when they should have been watching living flesh. "All laws of physics have been deduced from experiments made on dead matter," wrote Kervran.

Kervran's book explains the basics of his theories.

Living tissue, he went on to explain, always carries out a process which he named "biological transmutation." The process is so simple, so basic, that Louis Kervran didn't bother to explain how or why it happens. It just happens. Biological transmutation can go in either of two directions:

• Sometimes two atoms of different kinds combine, forming a larger, third kind of atom. This is nuclear fusion. Years later other people coined a term that describes it well: "cold fusion."

• Other times a large atom will split into two smaller atoms, each of different kinds. This is nuclear fission.

The box overleaf ("A Guide: Transmuting One Element Into Another") gives some technical details.

Physicists have never seen nuclear fusion or nuclear fission occur in living creatures. This, said Louis Kervran, is because they never looked. Louis Kervran looked, and here are just a few of the things he says he saw.

Chickens produce the calcium in their eggshells by transmuting potassium into calcium. A pig's intestines transmute nitrogen into carbon and oxygen. Cabbage transmutes oxygen into sulfur. Peaches transmute iron into copper.

In a report called "Non-Zero Balance of Calcium, Phosphorous and Copper in the Lobster," published in 1969, Kervran explained how lobsters perform nuclear fusion.

For his discovery of amazing things inside living flesh, Louis Kervran won the 1993 Ig Nobel Prize in the field of Physics.

The winner could not, or would not, attend the Ig Nobel Prize Ceremony.

IG NOBEL PRIZES A GUIDE: TRANSMUTING ONE ELEMENT INTO ANOTHER

How exactly does biological transmutation work? It's apparently simple to see, and as Louis Kervran wrote: "There is no chemistry involved."

When you look at a periodic table of the elements, you see that every element has a so-called "atomic number." The atomic number is the number of protons in the atom's nucleus. Here are a few elements, with their atomic numbers:

HYDROGEN – atomic number 1
SODIUM – atomic number 11
OXYGEN – atomic number 8
POTASSIUM – atomic number 19
CALCIUM – atomic number 20

The atomic numbers are the key to what is possible and what is not. Here are some examples, taken from Kervran's book, of how these elements transform, one kind becoming another.

- a SODIUM atom combines with an OXYGEN atom, and so becomes a POTASSIUM atom. [11 + 8 becomes 19]
- a CALCIUM atom splits into two pieces, a HYDROGEN atom and a POTASSIUM atom. [20 - 1 becomes 19]
- a POTASSIUM atom combines with a HYDROGEN atom, and so becomes a CALCIUM atom [19 + 1 becomes 20]

Kervran pointed out that "A law now emerges from these biological transformations: reactions at the nuclear level of the atom always involve hydrogen and oxygen."

He wrote that "it is usually vain to try to produce an element with biological transmutation if that element is not already present. In other words, what should be sought is the increase of an element (which always leads to the diminution of another), not its appearance from zero."

It should be noted that chemists and physicists all say they have never seen any of this happen, ever. Louis Kervran concluded that chemists and physicists are ignorant.

Mini-Dinosaurs, Mini-Princesses

"The miniman had a size like that of the Recent small ant and probably dwelt in caves after some development, or lived in simple type of houses constructed of plates of calcite or something similar. Furthermore, they knew letters and how to make cement by baking of calcite, and how to make china.

–explanation accompanying a photograph on page 271 of *Original Report of the Okamura Fossil Laboratory*

THE OFFICIAL CITATION

THE IG NOBEL BIODIVERSITY PRIZE WAS AWARDED TO

Chonosuke Okamura of the Okamura Fossil Laboratory in Nagoya, Japan, for discovering the fossils of dinosaurs, horses, dragons, princesses, and more than 1,000 other extinct "mini-species," each of which is less than 1⁄100 of an inch in length.

His studies were published in the series *Original Report of the Okamura Fossil Laboratory*, published by the Okamura Fossil Laboratory in Nagoya, Japan during the 1970s and 1980s. Earle Spamer, a scientist based at the Academy of Natural Sciences, in Philadelphia, is the world's leading (and perhaps only) expert on Chonosuke Okamura. Spamer has written three articles attempting to explain Okamura's work. They were published in the *Annals of Improbable Research*, vol. 1, no. 4 (July/August, 1995); vol. 2, no. 4, (July/August, 1996); and vol. 6, no. 6 (November/December, 2000). Much of the description below is adapted from Spamer's reports.

When a Japanese scientist examined rocks under a microscope, he discovered evidence that all modern living creatures are descended from tiny organisms resembling, in all but size, the big ones we see today. He gave a name to these extinct ancestor species: he called them "mini-creatures."

Chonosuke Okamura was a paleontologist who specialized in fossils of the unglamorous sort – invertebrate and algal specimens ranging in age from the Ordovician Period to the Tertiary Period. He published a series of dry, unglamorous reports.

But everything changed with the publication of *Original Report of the Okamura Fossil Laboratory*, number XIII. There Okamura showed photographs of a perfectly preserved fossil duck from the Silurian strata of the Kitagami mountain range – a previously unknown species he called Archaeoanas japonica. Okamura's illustration (reproduced on p.180) clearly shows the specimen just as he

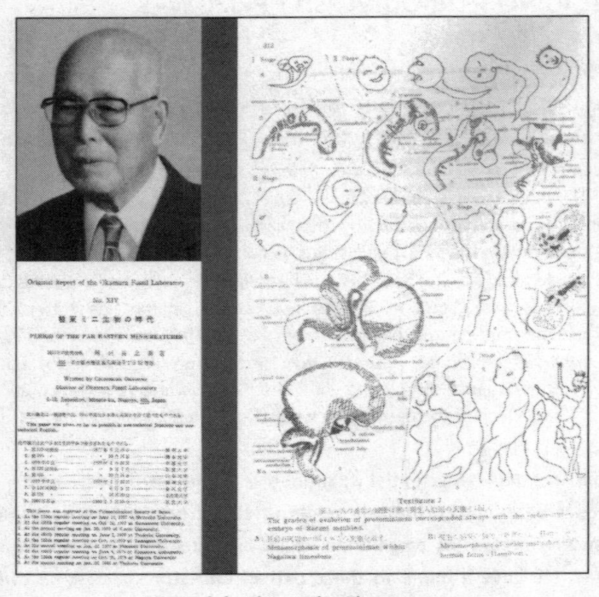

Top left: Chonosuke Okamura.
Bottom left: The title page from volume XIV of the *Original Reports of the Okamura Fossil Laboratory.*
Right: Stages in the evolution of mini-men into modern man. The main transition was a nearly thousandfold increase in height.

describes it: "in state of cramp through shock by being buried alive during the Silurian Period." The specimen measures just 9.2 mm long. This mini-duck is about the size of an aspirin tablet.

Okamura's subsequent reports were filled with remarkable photographs of the fossilized remains of all sorts of mini-creatures, each documented in photographs, with helpful diagrams and riveting descriptions that Okamura wrote himself in a helpful mixture of Japanese and broken English.

He describes mini-fishes, mini-reptiles, mini-amphibians, mini-birds, mini-mammals, and mini-plants. There are even mini-dragons, such as Fightingdraconus Miniorientalis and Twistdraconus miniorientalis.

Most of these newly discovered fossil taxa are subspecies of modern species. Okamora shows us the mini-lynx (Lynx lynx minilorientalis), the mini-gorilla (Gorilla gorilla minilorientalis), the mini-camel (Camelus dromedarius minilorientalus), the Silurian mini-snake (Y. y. minilorientalis), the mini-polar bear (Thalarctos maritimus minilorientalus), and the mini-common dog (Canis familiaris minilorientalis), whose "features were similar to those of a St. Bermard [sic] dog, but the length was only 0.5 mm."

Okamura also discovered the ancestral forms of extinct species; for example, a mini-pteradactyl (Pteradactylus spectabilis minilorientalus) and a children's perennial favorite, the mini-brontosaurus (Brontosaurus excelsus minilorientalus).

All of these creatures are less than a centimeter in length. Some are barely a millimeter long.

In most of the descriptions, Okamura combines scientific deduction with sympathetic observation. For example, in his description of Lynx lynx minilorientalis he notes:

"Some look frightened in anger against a sudden

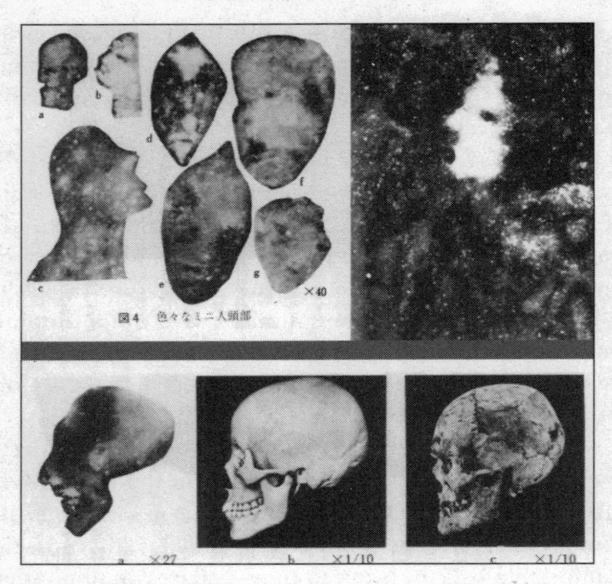

Fossils of mini-people, each of whom was approximately one millimeter tall. Upper left: Faces of mini-men. (Okamura cut these images from larger photographs, presumably to make them easier for non-specialists to recognize) Upper right: A specimen of Homo sapiens minilorientalis. Okamura wrote that she was a Nagaiwa mini-woman "about 30 years of age [and who] seems to be wearing a mantle of some kind on which many small dragons have been pasted, perhaps an after-death phenomenon." Lower panel: The skulls of a mini-man (left), a modern human (center) and an early hominid or proto-human (right). The mini-man skull is much smaller than the other two – an entire mini-man, from head to toe, was about as long as a modern infant's fingernail.

convulsion of nature while others are indifferent or even have sunk their heads on their breasts having lost the power of resistance. These are remnant remained forms of psychical movement showing the degree of development of intelligence."

Certainly, Okamura's greatest claim to fame is the discovery of the mini-man, Homo sapiens minilorientales. In a lengthy and meticulous anatomical discussion, illustrated with hundreds of photomicrographs, the earliest ancestors of humans are described. "The Nagaiwa miniman had a stature of only 1⁄350 that of the Recent man but with the same shape." The tools of these mini-people are described, too, including "one of the first metallic implements."

Okamura gives us keen insights into the lives of these mini-people. Consider these three of his observations:

- "All the women in Figure 70 have closed mouths and [are] seen to be undergoing pain by being buried alive in boiling mud, while the old woman in figure 1 has a wide open mouth, looking like one who has lost her senses."
- "In this photo, two totally-naked homos, facing each other, are moving their hands and feet harmoniously on good terms. We can think of no other scene than dancing in a present-day style."
- "They were polytheists and had many idols installed."

Okamura points out to us "the oldest hair styles;" "a quick-footed Nagaiwa mini-woman [who was] probably a hard worker;" a mini-woman who "seems to have been a person of noble rank;" and the "head of a miniman in the alimentary canal of a dragon."

The Nagaiwa mini-people were artisans, too, producing a broad variety of sculpture. "What may be regarded as the most elaborate piece of work," Okamura tells us, is that of a "full-length portrait of a woman sitting on the neck of a dragon," who "may be putting on a hat." Okamura "presumes this to be some kind of goddess," whose "mammae seem to be quite swollen and sagging a little."

The Nagaiwa mini-world was not idyllic. Okamura illustrates a "Close nestling protomini-man and protomini-

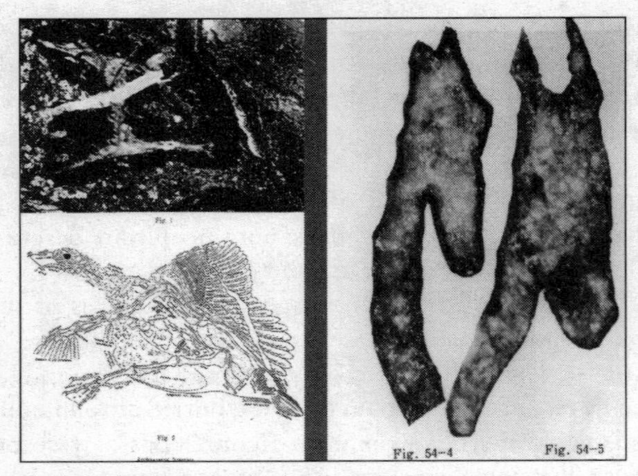

Upper left: A photomicrograph of the Silurian mini-duck, Archaeoanus japonica. It is just a few millimeters long. Lower left: Okamura's drawing identifies many of the body parts which some non-specialists might have difficulty recognizing. Right: A pair of mini-horses. Okamura wrote that "It is needless to say they had fertilized, grown and hatched in ovisacks to become the Equuus caballuus protominilorientalis. As it grew further, so it was destined to become the domestic animal." Each of these specimens is approximately 1.8 millimeters long, but presumably the species grows longer when the body fills out and it sprouts legs. (Okamura cut these images from larger photographs, presumably to make them easier for non-specialists to recognize.)

woman … both defying a dragon;" a "dragon strangling a girl;" and a "miniman offering a sacrifice to a brutal dragon", among other insightful tableaux. However, this review cannot even begin to place in proper perspective the wealth of detail described in Okamura's reports.

The relationship between mini-men and dragons appears to have been uncomfortable, if Okamura's interpretions are correct:

"From what the author could determine, the mini-men

lived in the ancient times having a high intellectual level with only flat nails for protecting themselves. Even if they grasped poles, using their free upper limbs, or used primitive metallic arms which seem to have existed, or hurled simply processed stones, it would have been most difficult to escape from the gluttoneous [sic] desire of countless flesh-eating hungry dragons."

The earlier forms of mini-people were without hands, but, Okamura tells us, "it would have made no difference if there had been a hand-to-hand fight with dragons, they still would have been defeated without the least resistance. The

IG NOBEL PRIZES

WHERE TO SEE OKAMURA'S PHOTOGRAPHS

Although Chonosuke Okamura sent copies of his *Original Report of the Okamura Fossil Laboratory* to many libraries around the world, many of these institutions seem not to have kept them. Earle Spamer, the leading expert on all things Okamura, obtained his copies at a yard sale. Spamer has compiled a list of some of the institutions that still hold copies of the Report:

Academy of Natural Sciences (Philadelphia)
Colorado School of Mines
Cornell University
Denver Public Library
Field Museum of Natural History
Harvard University, Museum of Comparative Zoology
Kent State University
Pell Marine Science Library (Narragansett, RI)
Smithsonian Institution
US Geological Survey (Reston, VA)
University of California at Los Angeles
University of California at San Diego
University of Houston
University of Texas at Austin
University of Wyoming

dragons would have mortally wounded them and crushed their bodies."

Okamura's startling observations were not totally devoid of emotion. He wrote, "The author will do his best to comfort their departed spirits."

The Okamura Fossil Laboratory has apparently not produced any new work since circa 1987. Chonosuke Okamura himself appears to have retired to a secluded life. His carefully detailed work simply drifted into obscurity, a sad example of what can happen when a scientist gets inadequate publicity.

For discovering tiny clues about our past, Chonosuke Okamura won the 1996 Ig Nobel Prize in the field of Biodiversity.

The winner could not, or would not, attend the Ig Nobel Prize Ceremony. The Ig Nobel Board of Governors tried, and failed, to find him.

The Remembrance of Water Passed

"Benveniste argues that the science establishment is inherently resistant to new ideas. 'Orthodox people are determined to block anything new in biology,' he says."

–from a news article in *Nature*

THE OFFICIAL CITATION

THE IG NOBEL CHEMISTRY PRIZE WAS AWARDED – TWICE – TO

Jacques Benveniste, prolific proseletizer and dedicated correspondent of *Nature*, for his persistent discoveries that:

• water, H_2O, is an intelligent liquid, and for demonstrating to his satisfaction that water is able to remember events long after all trace of those events has vanished; and that

• not only does water have memory, but the information can be transmitted over telephone lines and the Internet.

Jacques Benveniste's original study was published in the journal *Nature* in 1988 ("Human Basophil Degranulation Triggered by Very Dilute Antiserum Against IgE," *Nature*, vol. 333, no. 6176, June 30, 1988, pp. 816–18), but was later withdrawn at the insistence of the editors. His telephonic study was published as "Transatlantic Transfer of Digitized Antigen Signal by Telephone Link," J. Benveniste, P. Jurgens, W. Hsueh and J. Aissa, *Journal of Allergy and Clinical Immunology* – "Program and Abstracts of Papers to be Presented During Scientific Sessions AAAAI/AAI.CIS Joint Meeting," February 21-26, 1997.

Jacques Benveniste is the only person who has two Ig Nobel Prizes. He was honored for his discoveries – his memorable and repeated discoveries – that water (H_2O) has abilities that nobody had noticed.

In 1988, Benveniste, till that time a respected biologist at the highly respected INSERM (Institut National de la Santé et de la Recherche Médicale) in Paris, published a research paper in the respected journal *Nature*. In high-grade, professionally-obtuse technical language, the paper said (a) that water remembers things and (b) that Jacques Benveniste had proved it.

What's more, Benveniste told anyone who asked, his discovery explained how homeopathic medicines work.

Homeopathic drugs are, in effect, drugs that have had all the drug removed. Most scientists believe they don't work at all, except to the extent that people want to *believe* they work. After all, most good doctors and good scientists say bluntly, it's always been true that, for the most part, medicine consists of entertaining the patient whilst nature effects a cure.

Benveniste has been conducting his experiments for decades. Here's what he does. To a glass full of water he adds some particular chemical. Then he dilutes the mixture, then dilutes it again, then dilutes it again, then again, continuing until he has a glass full of nothing but absolutely pristine water. (You do much the same every day when you wash a glass with soapy water, then rinse it repeatedly till there's no soap left.) The pristine water in the glass, says Jacques Benveniste, remembers what the old water molecules told it: that once there was another chemical in that glass.

Benveniste's 1988 paper in *Nature* caused a first-class hoo-ha. To most scientists, the notion that "water has a memory" sounded nonsensical. But it was a big new idea, and scientists love nothing better than to try out the latest big new idea. And so thousands of scientists around the world did try. And, except for a few ardent advocates of homeopathic medicine, none of them could make it work. After wasting their time, some were disgusted, others amused. The biology magazine *The Scientist* reported that:

"Some scientists have had the good sense to turn to wit

instead of spleen ... Take NIH's Henry Metzger who tried unsuccessfully to duplicate Benveniste's finding that water retains a 'memory' of molecules it once contained. 'It's a shame,' Metzger sighs. 'It still takes a full teaspoon of sugar to sweeten our tea.'"

In 1991, for his insight that water has memory, Jacques Benveniste won the very first Ig Nobel Prize in the field of Chemistry. Not long afterwards, Linus Pauling, the only man who won two undivided Nobel Prizes (one for Chemistry, one for Peace), told the Ig Nobel Board of Governors that he hoped Edward Teller, winner of that year's Ig Nobel Peace Prize (see the section of this book on Edward Teller, "Father of the Bomb") would become the first man to have two Ig Nobel Prizes. But Pauling's hope was not realized.

Benveniste kept doing experiments, publishing papers (generally in obscure places), and ridiculing those who questioned his claims.

Eventually he left INSERM (press reports were vague about whether he left by choice) and moved to his own company, Digital Biology Laboratory, which he says will one day be bigger than Microsoft.

At Digital Biology Laboratory, Benveniste has been working to become the new Thomas Edison and the new Bill Gates combined. As Edison recorded memories from people, Benveniste is recording memories from water. Once he has these memories in digital format, they can be transmitted over telephone lines or over the Internet. Soon, according to Benveniste, pharmacists will stop selling drugs that are pills or liquids. Instead, prescriptions will be filled by, essentially, connecting a telephone line to your glass of water. Digital Biology Laboratory expects to be the leader of an entirely transformed pharmaceutical industry, and so to make Jacques Benveniste very, very wealthy.

Transatlantic Transfer of Digitized Antigen Signal by Telephone Link. *J. Benveniste, P. Jurgens, W. Hsueh and J. Aissa.* Digital Biology Laboratory (DBL), 32 rue des Carnets, 92140 Clamart, France and Northwestern University Medical School, Chicago, IL 60614, USA.

Ligands so dilute that no molecule remained still retained biological activity which could be abolished by magnetic fields [1-3], suggesting the electromagnetic (EM) nature of the molecular signal. This was confirmed by the electronic transfer to water (W) of molecular activity, directly or after computer storage [4-7]. Here, we report its telephonic transfer. Ovalbumin (Ova), or W as control,

The report that won Jacques Benveniste his second Ig Nobel Prize. Benveniste and his colleagues report that they collected some memories from a beaker of water, then sent those memories electronically over a telephone line. At the other end of the phone line, the memories were played through a loudspeaker into a beaker of water for 20 minutes. The water from this second beaker was then perfused into the heart of a dead guinea pig. The guinea pig heart reacted exactly as it would if the second beaker of water remembered the same things as the first beaker of water remembered. (Note: some observers find these concepts a little difficult to comprehend.)

In 1997, Benveniste filed a lawsuit against three prominent French scientists, two of them Nobel Laureates, who publicly expressed doubt about his work. The lawsuit was thrown out of court in 1988. Later that year, Jacques Benveniste became the first person to be awarded a second Ig Nobel Prize. This time around, he was honored for his discovery that watery memories can be transmitted by telephone and over the Internet.

The winner could not, or would not, attend the Ig Nobel Prize Ceremony, either in 1991 or in 1998. At the 1998 Ceremony, both the magician James Randi and the chemist Dudley Herschbach (see box overleaf) gave personal tributes to Benveniste.

At his laboratory in France, Benveniste told a reporter from *Nature* magazine he was "happy to receive a second Ig Nobel Prize, because it shows that those making the awards don't understand anything. People don't give out Nobel Prizes without first trying to find out what the recipients are doing. But the people who give out Ig Nobels don't even bother to inquire about the work."

The *Nature* report concludes with this paragraph:

"Harvard chemist Dudley Herschbach, who won the 1986 Nobel Prize in Chemistry, finds Benveniste's claims 'very hard to reconcile with what we know about molecules.' Herschbach considers the second 'Ig' prize 'very well deserved.' And he just might win a third one if he keeps going in this way."

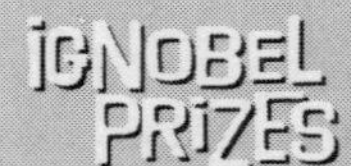

A FLOWING PERSONAL TRIBUTE

Here is Nobel Laureate Dudley Herschbach's moving tribute to two-time Ig winner Jacques Benveniste, delivered at the 1998 Ig Nobel Prize Ceremony.

"Immortal science, like great art, opens new perspectives on nature. Jacques Benveniste did so in 1988, when he published an astonishing article in *Nature* magazine. It reported his conclusion that water, once having encountered a biologically active molecule, remembered the experience very well – so well that long afterwards it could transmit the characteristic biological activity. The resonance with another classic of French literature – Proust's *Remembrance of Things Passed* [sic] – was uncanny.

"I must admit I was initially skeptical of his new work. It seemed quite incredible that specific biological activity could be transmitted by telephone or by the Internet. But Benveniste reports he has done thousands of experiments, simply by recording in the ordinary audio range signals from water. He does emphasize that, for his method to work, the water must have been 'informed' by receiving vibrations from the appropriate biomolecule. I've read several reports from Benveniste's lav–laboratory, which is named the Digital Biology Lavatory– er, Laboratory.

"The results led me to try some similar experiments with vibrating water – water that I was certain had indeed been informed of biological activity and might have remembrance of things passed. I have recorded these experiments, and now transmit them to you. [At this point, Professor Herschbach played a tape recording of a toilet being flushed.] I trust you heard that.

"These experiments, which you can easily replicate, indicate that although Benveniste's remarkable work may not imitate nature, it certainly offers a new perspective on the call of nature."

Discoveries – Things that Fall or Rise

The fall of mankind, perhaps from toilets, raises many questions. So does the fall of certain objects and the rise of others. Here are four investigations of things that lifted or plunged:

- Injuries Due to Falling Coconuts
- The Fall of Buttered Toast
- The Collapse of Toilets in Glasgow
- Levitating Frogs

Injuries Due to Falling Coconuts

"Falling coconuts can cause severe injuries. Many coastal villages in the tropical Pacific are surrounded by tall coconut palms. In this paper, four head injuries which resulted from falling coconuts in New Guinea are described. The physical forces involved in being struck by a falling coconut are discussed."

–from Peter Barss's report

THE OFFICIAL CITATION

THE IG NOBEL MEDICINE PRIZE WAS AWARDED TO

Peter Barss of McGill University, for his impactful medical report "Injuries Due to Falling Coconuts."

His study was published in *The Journal of Trauma*, vol. 21, no. 11, 1984, pp. 90–1.

As a young Canadian doctor newly arrived in Papua New Guinea, Peter Barss wondered what were the most common kinds of injuries that brought people to the Provincial Hospital in Alotau, Milne Bay Province. A surprisingly high percentage of these injuries, he discovered, were due to falling coconuts.

A relatively small number of people are actually killed by falling coconuts. One was the case of:

"a man who had come down to visit the coast from his home in the mountains of the island where there are few palm trees. He was perhaps unaware of the dangers of falling coconuts. He was standing beneath a tree as another man kicked down a coconut. It struck him squarely on the top of his skull; he dropped, and died within a few minutes."

Coconut palms, Dr Barss points out, grow to a great height. This is particularly true of the Cocos nucifera, variety typica, the most common type in Milne Bay Province.

"The trees grow continuously in height for 80 to 100 years, commonly reach 24 to 30 meters, and can be as high as 35 meters. The coconuts are attached high up in bunches at the top of the trunk ... They are sometimes harvested green for drinking, which is done by climbing the tree and cutting, kicking, or pulling loose the coconuts. [Dry coconuts] sometimes fall during heavy wind or during prolonged rainfall when the weight of the husks may increase. Houses are often built close to coconut palms. It is not surprising that adults or children are occasionally struck by falling nuts."

Dr Barss's report includes the first thoroughgoing technical analysis of the unfettered descent of a coconut (see overleaf). Important and interesting though the physics may be, the report's greatest import, like that of a falling coconut, is on the health of ordinary people. Dr Barss's hard-hitting conclusion says:

"The physical forces involved in a direct blow to the skull by a falling coconut are potentially very large. Glancing blows will, of course, be less serious. It seems unwise to locate dwellings near coconut palms, and children should not be allowed to play under coconut trees with mature nuts."

Dr Barss has, in his career, treated many kinds of injuries, some peculiar to the parts of the world in which he was living at the time. Visitors to what are, for them, far-off climes would do well to consult the medical literature for a quick head's up on what to watch out for when they arrive. Dr Barss has published more than 40 medical reports including several about illnesses and injuries characteristic of the south Pacific. These include: "Injuries Caused by Pigs in Papua New Guinea" (*Medical Journal of Australia*, vol. 149, December 5-19, 1988, pp. 649–56); "Grass-Skirt Burns in Papua New Guinea" (*The Lancet*, 1983); "Penetrating

Peter Barss terminates his acceptance speech at the behest of Miss Sweetie Poo. Photo: Diana Kudarayova/*Annals of Improbable Research*.

Wounds Caused by Needle-Fish in Oceania" (*Medical Journal of Australia*, 1985); and "Inhalation Hazards of Tropical 'Pea Shooters'" (*Papua and New Guinea Medical Journal*, 1985).

All of these are real and present dangers. But it is for looking into the matter of injuries due to falling coconuts – and for doing something about them – that Peter Barss won the 2001 Ig Nobel Prize in the field of Medicine.

Dr Barss traveled, at his own expense, from Montreal to the Ig Nobel Prize Ceremony. In accepting the Prize, he showed slides and said:

"I did this work in Papua New Guinea. I brought a few pictures of the wonderful people I worked with that helped me do my research. These are the type of trees people fall from ... and this is a man with a spinal cord injury falling from a tree being taken away ... Most of these people die, unfortunately. This is a simple device for removing breadfruit from a tree, to prevent injuries ... This one is just a simple prevention measure of pruning mango trees so you don't have to climb so high and fall so hard. Some of the heights of tropical trees are about the same as 10-storey buildings, so the mass of a falling coconut is about a metric ton with a direct hit. So, the worst place to be when a coconut falls, is asleep under the tree. Because your head is on the ground and you have a zero, uh, stopping distance so the physicists know the kinetic energy is infinite. It's better to be standing up and get knocked down ..." (At this point, Miss Sweetie Poo terminated Dr Barss's speech.)

IG NOBEL PRIZES TECHNICAL ANALYSIS OF A FALLING COCONUT

Here is Peter Barss's description of the Newtonian mechanics of a free-falling coconut:

"An average unhusked, mature dry coconut may weigh from 1 to more than 2 kilograms. A nut whose husk is soaked with water, or a green coconut, can weigh as much as 4 kilograms. When such a mass is accelerated by gravity, after falling from a height approximately equal to a 10-storey building, and then comes to rest by being suddenly decelerated onto someone's head, it is not surprising that severe head injuries sometimes occur. If a coconut weighing 2 kilograms falls 25 meters onto a person's head, the impact velocity is 80 kilometers/hour. The decelerating force on the head will vary depending on whether a direct or glancing blow is received. The distance in which the coconut is decelerated is also an important factor. Thus an infant's head lying on the ground would receive a much greater force than that received by the head of a standing adult, that dropped as it was struck. For a stopping distance of 5 centimeters and a direct blow, the force would be 1,000 kilograms."

The Fall of Buttered Toast

"We investigate the dynamics of toast tumbling from a table to the floor. Popular opinion is that the final state is usually butter-side down, and constitutes prima-facie evidence of Murphy's Law ('If it can go wrong, it will'). The orthodox view, in contrast, is that the phenomenon is essentially random, with a 50/50 split of possible outcomes. We show that toast does indeed have an inherent tendency to land butter-side down for a wide range of conditions. Furthermore, we show that this outcome is ultimately ascribable to the values of the fundamental constants. As such, this manifestation of Murphy's Law appears to be an ineluctable feature of our universe."

–from Robert Matthews's report

THE OFFICIAL CITATION

THE IG NOBEL PHYSICS PRIZE WAS AWARDED TO

Robert Matthews of Aston University, England, for his studies of Murphy's Law, and especially for demonstrating that toast often falls on the buttered side.

His study was published as "Tumbling toast, Murphy's Law and the Fundamental Constants," *European Journal of Physics*, vol.16, no.4, July 18, 1995, pp. 172–6. Details of the later empirical test were published in *School Science Review*, vol. 83, 2001, pp. 23–8.

The fall of buttered toast is, among other things, an old joke. In 1844, the poet and satirist James Payn wrote:

"I've never had a piece of toast
particularly long and wide,
but fell upon a sanded floor,
and always on the buttered-side."

More than a century later someone (there is much dispute as to who) pointed out that if cats always land on their feet, you could strap buttered toast onto the back of a cat, and the combination might spin

forever, suspended inches above the ground.

In 1995, Robert Matthews strapped mathematics onto the buttered toast question, and out dropped a revelation.

Matthews is a Chartered Physicist, a Fellow of the Royal Astronomical Society, and a Fellow of the Royal Statistical Society. He is a student of Murphy's Law. He took the toast question seriously.

There are many factors that must be considered. Matthews began by demolishing a cherished assumption:

"There is a widespread belief," he wrote, "that it is the result of a genuine physical asymmetry induced by one side of the toast being buttered ... This explanation cannot be correct. The mass of butter added to toast (on the order of 4 grams) is small compared to the mass of the typical slice of toast (on the order of 35 grams), is thinly spread, and passes into the body of the toast. Its contribution to the total moment of inertia of the toast – and thus its effect on the toast's rotational dynamics – is thus negligible."

Then, in a mere five pages of computation, Matthews explored the behavior of a rigid, rough, homogeneous rectangular lamina, mass *m*, side 2*a*, falling from a rigid platform set a height *h* above the ground. He considered the dynamics of this body from an initial state where this center of gravity overhangs the table by a distance *delta sub-nought*, and analyzed it mercilessly through all stages of its perilous journey to the final resting position at height *h* equals zero.

When he had finished, there was a startling insight:

"The formula giving the maximum height of humans turns out to contain three so-called 'fundamental constants of the universe.' The first – the electromagnetic fine-structure constant – determines the strength of the chemical bonds in the skull, while the second – the gravitational fine-structure constant – determines the strength of gravity. Finally, the so-called 'Bohr radius' dictates the size of atoms making up the body. The precise values of these three fundamental constants were built into the very design of the universe just moments after the Big Bang. In other words, toast falling off the breakfast table lands butter-side down because the universe is made that way."

This, of course, did not end the controversy – Murphy's Law forbids that ever happening. After Robert Matthews published his paper, other scientists leaped yackettingly into the arena. They quibbled furiously about parametric values, the calculus of variations, and certain fine points of stochastic estimation methodology. No matter. Matthews had produced a standard against which all crumby researchers, now and forever, must measure their work.

For adding a thick dollop of mathematics to a smear of butter and a slice of toast, Robert Matthews won the 1996 Ig Nobel Prize in the field of Physics.

The winner could not travel to the Ig Nobel Prize Ceremony, but instead sent an audio-taped acceptance speech. True to Murphy's Law, the tape arrived at Harvard four days after the Ceremony. In the speech, Dr Matthews said:

"Thank you very much for this award. Proving that Murphy's Law – if something can go wrong, it will – is built into the design of the universe has brought me, as one of the most pessimistic people on earth, a lot of pleasure, and so has this Ig Nobel Prize. There is, of course, a more serious side to my work, I just can't remember what it is. Oh, yes, I know. I should get out more."

Matthews thereafter continued his research both on matters covered by Murphy's Law and on other practical questions outside the Law. Why there are so many odd socks in our drawers; why rope or string so often seems to acquire knots; why places you're looking for so often lie in awkward places on maps; whether to take an umbrella following a forecast of rain; whether to switch queues while waiting in a supermarket – Robert Matthews has attacked all these and more with dash, vim and stylish mathematics.

In 2001, he returned to the question of buttered toast. Having already solved it theoretically, he now attacked the problem empirically. That is to say, he conducted an experiment:

"A total of just over 1,000 schoolchildren (70% primary, 30% secondary) from schools across the UK took part in the three experiments, performing a total of over 21,000 drops of toast. The dedication of some of the school teams was impressive, with 22 reporting at least 100 drops, 10 at least 400 drops, and two conducting over 1,000. The overall results of the three basic experiments were as follows:

"Out of a total of 9,821 drops, there were 6,101 butter-down landings, a rate of 62%, which is 12% higher than the 50% rate expected if – as many scientists have claimed – toast is as likely to land butter-up as down, and its final state is random."

And thus Robert Matthews demonstrated, both theoretically and experimentally, that nature abhors a newly vacuumed floor.

The Collapse of Toilets in Glasgow

"Three cases are presented of porcelain lavatory pans collapsing under body weight, producing wounds which required hospital treatment. Excessive age of the toilets was implicated as a causative factor. As many toilets get older, episodes of collapse may become more common, resulting in injuries."

–from Wyatt, McNaughton, and Tullet's report

THE OFFICIAL CITATION

THE IG NOBEL PUBLIC HEALTH PRIZE WAS AWARDED TO

Jonathan Wyatt, Gordon McNaughton, and William Tullet of Glasgow, for their alarming report, "The Collapse of Toilets in Glasgow."

Their study was published in the *Scottish Medical Journal*, vol. 38, 1993, p. 185.

Three doctors working in the Department of Accident and Emergency of the Western Infirmary in Glasgow noticed an unusual conjunction of events: "three patients who presented during a period of six months with injuries sustained whilst sitting on toilets which unexpectedly collapsed."

This, they decided, was worth investigating.

The three reports were, upon re-examination, remarkably dissimilar.

The first involved a 14-year-old girl who weighed 83 kilograms, who "sustained a 7-centimeter wound to the posterior aspect of her right thigh when she sat on a school toilet which promptly collapsed."

The second was a 34-year-old man who weighed 70 kilograms, who "sustained a 6-centimeter wound to his right buttock when a toilet collapsed under him during defecation."

The third patient was a 48-year-old man who weighed 76 kilograms, who was "sitting on a toilet which disintegrated, causing multiple wounds to both buttocks."

In all three cases, the collapse was of the porcelain lavatory pan, not of the toilet seats. The official report says that "the exact ages and origins of the pans were not known, but all were described as white porcelain."

In each case, the attending physician cleansed and sutured the wound under local anesthetic, and the patient went on to make a full recovery.

Drs Wyatt, McNaughton, and Tullet conclude their report with a grim, but not entirely discouraging, summary, and a practical suggestion:

"The toilet collapses described in this study, whilst not producing life-threatening injuries, resulted in considerable embarrassment and discomfort for those involved. Toilet collapse appears to be unusual; a literature search failed to reveal any previous similar reports. The cause remains unclear, except that all of the toilets were believed to be

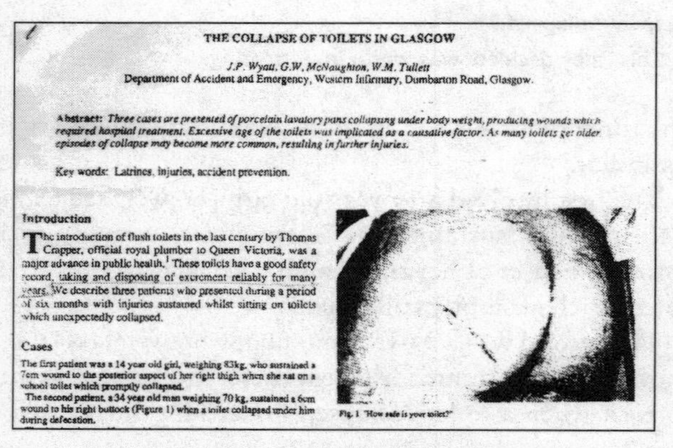

THE COLLAPSE OF TOILETS IN GLASGOW

J.P. Wyatt, G.W. McNaughton, W.M. Tullett
Department of Accident and Emergency, Western Infirmary, Dumbarton Road, Glasgow.

Abstract: *Three cases are presented of porcelain lavatory pans collapsing under body weight, producing wounds which required hospital treatment. Excessive age of the toilets was implicated as a causative factor. As many toilets get older episodes of collapse may become more common, resulting in further injuries.*

Key words: Latrines, injuries, accident prevention.

Introduction

The introduction of flush toilets in the last century by Thomas Crapper, official royal plumber to Queen Victoria, was a major advance in public health.[1] These toilets have a good safety record, taking and disposing of excrement reliably for many years. We describe three patients who presented during a period of six months with injuries sustained whilst sitting on toilets which unexpectedly collapsed.

Cases

The first patient was a 14 year old girl, weighing 83kg, who sustained a 7cm wound to the posterior aspect of her right thigh when she sat on a school toilet which promptly collapsed.

The second patient, a 34 year old man weighing 70 kg, sustained a 6cm wound to his right buttock (Figure 1) when a toilet collapsed under him during defecation.

Fig. 1 "How safe is your toilet?"

Wyatt, McNaughton, and Tullet's Prize-winning paper.

very old. We would therefore advise that the older porcelain familiar to so many of us should be treated with a certain degree of caution. An obvious way of using a toilet without fear of it collapsing is to take a continental approach and not to sit down, but to adopt a hovering stance above it."

For their contributions to the safety and peace of mind of the populace of greater Glasgow, Jonathan Wyatt, Gordon McNaughton, and William Tullet won the 2000 Ig Nobel Prize in the field of Public Health.

Jonathan Wyatt and Gordon McNaughton traveled, at their own expense, from Glasgow to the Ig Nobel Prize Ceremony. McNaughton wore a kilt for the occasion. Wyatt did not. In accepting the Prize, they kept their remarks short and to the point:

Jonathan Wyatt: "Thank you very much, ladies and gentlemen, it's a wonderful honor to be here tonight. You won't believe it, but our research has previously been dismissed as a mere flush in the pan. But this ceremony has been able to show it in its true light. Gordon – this is Gordon here, from Scotland – would particularly like to thank you for your American hospitality. Just to explain, for those of you who don't understand why it is that Scotsman wear kilts, he's been trying out your toilets and he finds it very easy with this particular kilt to try out a number of toilets and, so far, so good. I'll hand you on to Gordon. Thank you very much."

Gordon McNaughton: "To link in with tonight's intelligence theme, we thought it would be an opportunity to mention one of England's most famous plumbers, and that is Mister Thomas Crapper, who, in fact, without his intelligence and his invention of the flush toilet, we would not be standing here today. Thank you very much."

Two months after the Ceremony, the Ig Nobel Board of

Gordon McNaughton, flushed with pride at the Ig Nobel Ceremony. Photo: AIR.

Governors received a note from a Mr Alasdair Baxter of Nottingham, England, which read as follows:

"Do pardon this intrusive e-mail but I have a sneaking feeling that I could be one of the victims of a collapsing toilet mentioned in the research paper published in the *Scottish Medical Journal*, vol. 38, 1993, p. 185. While working as a temporary schoolteacher in Coatbridge near Glasgow in August 1971, I sat on a toilet in the school which collapsed and caused severe lacerations of my buttocks and back. Sadly, I cannot get ready access to the *Scottish Medical Journal* to check it out and if you can send me a copy of the article, I shall be very grateful indeed. Alternatively, if you have an e-mail address for either Dr Wyatt or Dr McNaughton, I shall e-mail them directly. Thank you in anticipation."

The Board sent a copy of Mr Baxter's note to Dr McNaughton. Dr McNaughton replied with the bittersweet news that Mr Baxter's buttocks were not amongst those which he, Dr Wyatt, and Dr Tullet had had the honor of examining.

Levitating Frogs

"If the frog is initially in equilibrium, there are no forces on it. By changing shape (e.g. from a sphere to an ellipsoid) the induced moment will change (Landau et al., 1984), and the force will no longer be zero, so the frog will start to oscillate about a slightly different point. By repeating this manoeuvre at the frequency of oscillations in the minimum, the oscillations will be amplified by parametric resonance until the frog leaves the stable zone. [But] this is a tiny effect, because the shape-dependence of *m* is of the order of 10 to the minus fifth power, so escape would require 1,000,000 such 'swimming strokes'; therefore the frog would have to be persistent as well as highly coordinated."

–from Geim and Berry's report

THE OFFICIAL CITATION

THE IG NOBEL PHYSICS PRIZE WAS AWARDED TO

Andre Geim of the University of Nijmegen, The Netherlands, and Sir Michael Berry of Bristol University, England, for using magnets to levitate a frog.

Their study was published as "Of Flying Frogs and Levitrons," *European Journal of Physics*, vol. 18, 1997, pp. 307–13. Andre Geim's web site (www.hfml.sci.kun.nl/froglev.html) has short videos of the frog, a cricket, a strawberry and a drop of water being levitated.

"No, you cannot magnetize a frog," is the conclusion most scientists would have reached had they ever given a thought to the question, "Can you magnetize a frog?" Most scientists should consider themselves fortunate, because had they considered the question, and reached that conclusion, and publicly staked their reputation on saying so, they would have been dead wrong.

The frog levitation was a one-man *tour de force*. The theoretical underpinnings were a joint effort. Michael Berry explains:

"The flying frog was Andre Geim's experiment. I was told about it after giving a lecture on the physics of the levitron – a toy in which a magnetized spinning-top floats above a magnetized base. It seemed that the flying frog and the floating top ought to depend on similar physical principles, so I got in touch with Andre. Then we worked together, to extend to the frog the explanation I had previously found for the levitron.

"It is surprising at first to see the frog suspended in midair, in apparent defiance of gravity. It is supported by the force of magnetism. The force comes from a powerful electromagnet. It is able to push upwards on the frog because the frog is a magnet, too, albeit a weak one. The frog is intrinsically non-magnetic but becomes magnetized by the field of the electromagnet – this is called 'induced diamagnetism.' Most substances are diamagnetic, and Andre was able to levitate a variety of objects, including drops of water and hazelnuts.

"In principle, a person could be magnetically levitated too – like frogs, we are mostly water. The field would not have to be stronger, but would have to fill the much larger volume of a person, and that has not been achieved yet. I have no reason to believe such levitation would be a harmful or painful experience, but of course nobody can be sure of this.

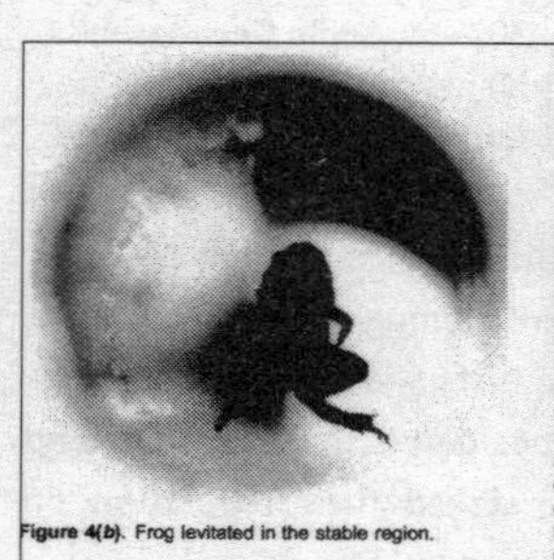

Figure 4(*b*). Frog levitated in the stable region.

A variety of diamagnetic objects was inserted into the magnet, and the current through the coils adjusted until stable levitation occurred (figure 4(*b*)). The corresponding fields B_0 were all close to the calculated 16 T, and the objects always floated near the top of the

Of course this represe
currents localized in
charge, so the living
Indeed, they emerged
without suffering any
also Schenck (1992) a
As we showed ea
paramagnets stably.
be achieved, and fron
clear that this occurs
of the solenoid—rathe
$\chi_{paramagnetic} \approx 10^{-3}$
vertically stable but
some paramagnetic ob
stainless steel, param
were suspended in th
were held against the
a few occasions, para
contact, but were fou
current of paramagne
for example by coveri
gauze, the objects slip
against the wall.

6. Discussion

Our treatment of dian

Geim and Berry's Prize-winning report.

Nevertheless, I would enthusiastically volunteer to be the first levitatee.

"The tricky part of the physics is to understand why the equilibrium of the frog is stable – that is, why it remains suspended. Most physicists would – mistakenly – expect the frog to slip sideways out of the field, and fall (an analogy is the instability of a pencil balancing on its point). This wrong expectation is based on a theorem proved by Samuel Earnshaw in 1842: no stationary object can be held stably by magnetism and gravity alone. But the frog is not stationary. There is the circulation of electrons in the creature's atoms. These are small effects, but they mean that Earnshaw's theorem does not strictly apply, and this opens the possibility that the equilibrium can be stable.

"The trick is to get the forces to balance in these regions. If you get it wrong, the frog will fall."

Miss Sweetie Poo terminates the acceptance speech of flying frog researcher Andre Geim. Photo: Herbert Blankesteijn, Handelsblad NRC, used with permission.

For their magnetic levity, Andre Geim and Michael Berry won the 2000 Ig Nobel Prize in the field of Physics.

Andre Geim flew at his own expense from Nijmegen, The Netherlands, to the Ig Nobel Prize Ceremony. In accepting the Prize, he said:

"Our story contains some unappreciated knowledge about magnetism. We want to accept this prize also on behalf of the hundreds who wrote to us with their ideas. The enquiries came from engineers who wanted to use levitation for everything from waste recycling and materials processing, to levitating sports shoes and jewelry in shop windows; from our physicist colleagues, some of whom admitted that after learning about the frog they finally understood some of their old results; from chemists and biologists who did not want to wait for a space shuttle and realized they could do micro-gravity experiments in a magnet; from servicemen to pensioners and from prisoners to priests. Sometimes, their ideas were bright and unexpected, sometimes goofy, some-times ridiculous or even mad, but always creative. Even more rewarding were letters from children all over the world who wrote 'I am nine years old and want to become a scientist.' " (Dr Geim continued for a short while longer, but Miss Sweetie Poo, who herself was eight years old, terminated his speech.)

Troy & the Grizzly Bear

Of all the Ig Nobel Prize winners, one defies all efforts to categorize him. Troy Hurtubise rates a chapter entirely of his own. Here, in brief, is the story of:

- Troy & the Grizzly Bear

Troy & the Grizzly Bear

"Like Ahab before him, Troy Hurtubise obsessively stalks the Great Other, donning 147 pounds of homemade armor, suffering countless test-pummelings, and sliding into bankruptcy as he awaits the ultimate showdown."

–from a 1997 *Outside* magazine article about Troy Hurtubise

THE OFFICIAL CITATION

THE IG NOBEL SAFETY ENGINEERING PRIZE WAS AWARDED TO

Troy Hurtubise, of North Bay, Ontario, for developing, and personally testing a suit of armor that is impervious to grizzly bears.

Troy Hurtubise and his work are shown in the documentary film *Project Grizzly*, produced by the National Film Board of Canada. Further information and video clips are on Troy's web site www.projecttroy.com

At age 20, out alone panning for gold in the Canadian wilderness, Troy Hurtubise had an encounter of some sort with a grizzly bear. Troy has devoted the rest of his life to creating a grizzly-bear-proof suit of armor in which he could safely go and commune with that bear. The suit's basic design was influenced by the powerful humanoid-policeman-robot-from-the-future title character in *RoboCop,* a movie Troy happened to see shortly before he began his intensive research and development work.

Troy is a pure example of the lone inventor, in the tradition of James Watt, Thomas Edison, and Nikola Tesla. Regarded by some as a half-genius, by others as a half-crackpot, he has unsurpassed persistence and imagination. Troy also is very careful. The proof that he is very careful is that he is still alive.

A grizzly bear is tremendously, ferociously powerful. Troy realized that he would be wise to test his suit under controlled conditions prior to giving it the ultimate test. He spent seven years and, by his estimate, $150,000 Canadian, subjecting the suit to every large, sudden force he could devise. For almost all of the testing, Troy was locked inside the bulky suit, despite his being severely claustrophobic.

The suit is a technical wonder, especially when one realizes that Troy had to assemble it mostly from scrounged materials. Here are the technical specs:

NAME: Ursus Mark VI
MATERIALS:
- Fireproof rubber exterior (from Minnesota)
- Titanium outer plates (from Hamilton, Ontario)
- Suit joints made of chain mail (from France)
- Tek plastic inner shell (from Japan)
- Inner layer of air bags
- Duct tape

HEIGHT: 2.18 metres (7 feet, 2 inches), with head-top camera attachment.
WEIGHT: 66.68 kilograms (147 pounds).
HEADPIECE: Two-chamber headpiece. Inner helmet: specially modified Shoei motorcycle helmet. Outer chamber: aluminum/titanium alloy shell. Dimensions: approx. 60 centimeters (2 feet) deep; 45 centimeters (1 foot, 6 inches) wide.
COOLING SYSTEM: Battery-powered twin-fan ventilation system draws cool air into helmet and vents warm air.
RADIO SYSTEM: Voice-activated two-way radio.
VIEWING SYSTEM: Helmet-mounted miniature camera with wide-angle viewscreen.
BLACK BOX: Voice-activated recording device located on the rear-right side of head piece, to record bear sounds or, in the event of a catastrophic failure of the Ursus Mark VI, last words.

DEFENSIVE SYSTEM: Trigger-finger-activated "blaster can" on right arm, capable of spraying a 38-centimeter (15-inches) diameter cone of bear repellant for a distance of 4.6 meters (15 feet), for a duration of 7 seconds.
BITE BAR: Pressure-sensitive strip located on right arm, to measure the biting power of a grizzly.

TESTING ON SUIT:
• TRUCK: 18 collisions with a three-ton truck travelling at 50 kilometers an hour (30 mph).
• RIFLE: Shot at with 12-gauge shotgun, using "Sabot" slugs.
• ARROWS: Armor-piercing arrows, fired from 45-kilogram (100-pound) bow.
• TREE TRUNK: Two collisions with a 136-kilogram (300-pound) tree from a height of 9 meters (30 feet).
• BIKERS: Assault by three bikers – the largest, 2.05 meters (6 feet, 9 inches) tall, weighing 175 kilograms (385 pounds). Biker armaments: splitting ax, planks, baseball bat.
• ESCARPMENT: Jumped off cliff, falling more than 15.25 meters (150 feet).

The Ursus Mark VI has some minor drawbacks which will be remedied before Troy considers the design to be optimal. Eventually it will be possible to take more than five steps without falling over if the ground is not perfectly level.

Future versions of the suit will be lighter and more flexible.

Troy is a natural leader blessed with boundless charisma, charm, and good humor. He works with a team of volunteers, who are ever at the ready to drop whatever they are supposed to be doing and help Troy build and test each new

Troy Hurtubise in his grizzly-bear-proof suit of armor. Photo: Greg Pacek, courtesy of the National Film Board of Canada.

version of his invention. They also help Troy videotape most of the tests. Some of the best early footage was included in the National Film Board of Canada's 1997 documentary *Project Grizzly*. *Project Grizzly* also shows Troy's first sally back into the wilderness, on horseback and with the proper clothing, in search of the bear. The producers promoted the film with a cheery invitation:

"Join Troy as he tests his armour and courage, in stunts that are both hair-raising and hilarious. Journey with this modern-day Don Quixote and his band of men, as they travel from the donut shops and biker bars of North Bay to the mythic Rocky Mountains, for a date with destiny."

Troy enjoyed the attention, but was disappointed that the tone of the movie does not properly highlight his commitment to doing research on the science of grizzly bears.

The film does make abundantly clear that Troy does not let setbacks discourage him. One setback came in the late 1990s, when Troy was forced to declare bankruptcy, and the Ontario bankruptcy court took possession of his suit. Since that time, the court has been trying to find a buyer. The court occasionally gives Troy permission to borrow the suit, most commonly for television interviews and other public appearances when a buyer might be likely to see and decide to covet it.

For conceiving of, and building, and testing the suit, and for keeping it and himself intact the whole while, Troy Hurtubise won the 1998 Ig Nobel Prize in the field of Safety Engineering. The Ontario bankruptcy court gave permission for the suit to accompany Troy to the Ceremony at Harvard.

Troy traveled from his home in North Bay, Ontario, to the Ig Nobel Prize Ceremony accompanied by his wife, Laurie, and a mysterious man named Brock, who wore a dark business suit, and whom Troy introduced as a court-appointed guardian for the suit, but who described himself as "the president of one of Troy's companies." The three of them had a minor, but not lengthy, adventure getting the suit through American customs inspection at Boston's Logan Airport. From there it was a short trip through Boston traffic, over the Charles River into Cambridge, and finally to Sanders Theatre at Harvard.

In accepting the Prize, Troy said:

"I'm still alive, anyway. What can I say? I'm just a simple man from northern Ontario, standing in the hallowed halls of Harvard. I say, feel the tension, man, what a ride! We must look past the obvious absurdity of some inventions and discoveries, and set aside the narrow-mindedness of science, if for but a moment, to allow the view to become unobstructed, with a vehicle called imagination.

"The Mark VI suit is bulletproof and fireproof, and all those kinds of nice things. The exoskeleton is pure titanium. The outside rubber base protects the electronics. For two years I had a problem of getting them to bond. So to bond the rubber to the titanium, I coated the inside of the suit – which you can't see – with 7,630 feet of duct tape.

"Tomorrow, at the Science Center at Harvard, I will unveil, as a world exclusive, the next prototype – the G-Man Genesis – and the science behind it."

The next day, Troy revealed – for the first time anywhere, and before an eager press of reporters and technologists – his plans for the next-generation suit. Estimated to cost $1.5 million in its first incarnation, the G-Man Genesis, he said, will be lighter, stronger, and far more maneuverable than its predecessor. A person wearing the suit will be able to run at full tilt, and to explore inside volcanoes.

Nobel Laureates (left to right) Glashow, Herschbach, Roberts, and Lipscomb cheer as Troy Hurtubise displays the Ig Nobel Prize he won for creating and personally testing a grizzly-bear-proof suit of armor. Referee John Barrett, wearing an inferior suit of armor, looks on in envy. Photo: Relena Erskine/Anna Boysen/AIR.

That evening, Troy appeared at two special public screenings of *Project Grizzly* at the Harvard's Carpenter Center before sold-out, adoring crowds.

Troy returned to Harvard the following year, to help honor the new crop of Ig Nobel Prize winners, and also to lecture at MIT, where he brought a roomful of engineers to a near frenzy of inspiration.

Since then, Troy has continued to do advanced research and development work, and to have unexpected adventures involving, among other things, NASA, the national Hockey League, an invention to separate oil from sand, a tapped phone, a mysterious nocturnal break-in, getting kicked in the crotch on television by comedian Rosanne Barr, a visit from al-Qaeda hijackers, and an encounter in a locked room with two Kodiak bears.

Tributes to Troy

Here are a few of the tributes given to Troy Hurtubise at the 1998 Ig Nobel Prize Ceremony, when Troy received his Ig Nobel Prize.

Tribute to Troy: Raw Power

by Colin Gillen, researcher, Wildlife Clinic, Tufts University Veterinary School of Medicine

"I've seen the aftermath of a hungry grizzly peeling the back end of a Winnebago off just like a can of sardines – just to get at an Oreo cookie. I've also seen the metal bear-proof food boxes used by our own US Forest Service in the back country rolled at least a half-mile through the trees and down the valleys and up the other hillside. They generally weigh at least 200 pounds, maybe 300 pounds with food in them. And I've witnessed a grizzly pick up a 200-pound hind quarter, unattached, of a bison in its teeth, and walk away without leaving a drag mark in the snow.

"I hold an intrigued sort of admiration for Troy and the unique niche that he's carved for himself in the field of bear research and engineering. Anyone seeing Troy's video will be interested, as I am, in the results of his actual field trials.

I wish him success and all the luck he can muster in his pursuits, and trust that all his risks are calculated in his favor."

Tribute to Troy: Materials Science

by Robert Rose, Professor of Materials Science, Massachusetts Institute of Technology

"Troy James Hurtubise, we recognize today your innovative use of modern materials technology in your field experiments on musculo-skeletal biomechanics, particularly in shock absorption.

"The impact of your research methodology will be felt in many areas. The more extensive and intimate involvement in research of human subjects, who after all are the only experimental animals with health insurance. The practice of field testing with baseball bats, which I would recommend for many university research projects I'm acquainted with. The secrets of hibernation would be of great practical significance to many of our students and faculty."

Tribute to Troy: Man and Bear

by Dudley Herschbach, Nobel Laureate, Professor of Chemistry, Harvard University

"I once encountered a grizzly bear. He was about as big as a house. Fortunately, he didn't like me, either."

Special Announcement About Bear Suits

by prominent New York customs attorney William J. Maloney

"I speak to you tonight on behalf of the Amalgamated Grizzly Bear Suit Workers of America, a union that has been proudly representing American workers in the protective grizzly bear suit industry for over 75 years. We have reason to believe that Ig Nobel Laureate Troy Hurtubise is manufacturing his suits in Canada with cheap, non-union labor. Furthermore, we believe that he is preparing to mass-produce these suits to flood the American market, undercutting sales of the higher quality grizzly suits manufactured by unionized American labor."

Inventions

Most inventors, like most of their inventions draw little public attention. Here are four exceptions:

- The Most Inventive Salesman
- The Kitty and the Keyboard
- AutoVision
- Patenting the Wheel

The Most Inventive Salesman

"The Trim-Comb is simply a small plastic comb with a razor blade inside of it. It was a great little invention, a low-cost, high-margin item that made us lots of money. And how good were Trim-Comb haircuts? As good as the person doing the cutting."

–Ron Popeil, in his autobiography, *The Salesman of the Century* . . .

THE OFFICIAL CITATION

THE IG NOBEL CONSUMER ENGINEERING PRIZE WAS AWARDED TO

Ron Popeil, incessant inventor and perpetual pitchman of late night television, for redefining the industrial revolution with such devices as the Veg-O-Matic, the Pocket Fisherman, Mr Microphone, and the Inside-the-Shell Egg Scrambler.

Ron Popeil tells his story in the book *The Salesman of the Century: Inventing, Marketing, and Selling on TV: How I Did It and How You Can Too!*

For more than four decades, American television has been filled with commercials for strange, cheap inventions with curious names – the Buttoneer, the Pocket Fisherman, the Mince-O-Matic, the Inside-the-Shell Egg Scrambler, and nearly countless others.

Behind all of them lay a relentlessly glib, and simply relentless, salesman/inventor named Ron Popeil.

Civilization had produced many inventors of half-needed devices. Civilization had produced many over-energized salesmen. Ron Popeil (he pronounces it "poe-*peel*") is both. The combination is not terribly special, but Ron Popeil, the man, is. More than anyone else, he used television as a super-effective, super-irritating, yet unquestionably compelling way to goose up the process of selling junky little gadgets.

Ron Popeil was born to a clan of inventive, cantankerous salesmen.

Young Ron's father, the inventor-salesman S.J. Popeil, designed and built a thingy called the Chop-O-Matic, "a simple chopping device that cuts up vegetables, potatoes, and meats." S.J. Popeil endured travails for his Chop-O-Matic. He filed a lawsuit against his uncle, the inventor/salesman Nathan Morris, for devising and marketing a similar device called the Roto-Chop. Nathan Morris was a veteran of such battles, chiefly fought against his own brother, the inventor/salesman Al Morris. S.J. Popeil managed to reach a legal settlement with his uncle Nathan, thus freeing the Chop-O-Matic from legal limbo. (Ironically, many years later, S.J. Popeil lost a legal battle with the Swiss inventors of an earlier Chop-O-Matic-like machine called the Blitzhacker.)

Eventually, Ron figured out how to well and truly promote and sell his papa's Chop-O-Matic.

Armed with this heritage, these skills, and this experience, Ron went on to invent or acquire a truly bewildering array of little machines, and advertise, advertise, advertise them down people's throats.

No one exactly needed any of these products, but with their snazzy names the cheap little gizmos were somehow weirdly enticing. Each filled a different need that people *almost* believed they had.

Television ads, cheaply made and cheaply broadcast, were the key to success. Ron Popeil devised ways to inexpensively produce and broadcast his ads again and again, *ad nauseam*, day and night (especially late night, when it cost him almost nothing to get air time). The sales pitch drilled deep into the nervous system of anyone whose television set happened to be turned on.

Popeil's ads were engineering marvels, as were his product names. Snappy and attention-grabbing, each commercial had a high-pressure announcer who dispensed a stream of rat-a-tat patter, harping on the product name. Full of cloyingly artificial friendliness, the announcer repeated the pitch several times, always sweetening the deal with idiotic little extras. "But wait," he would say, "there's more!"

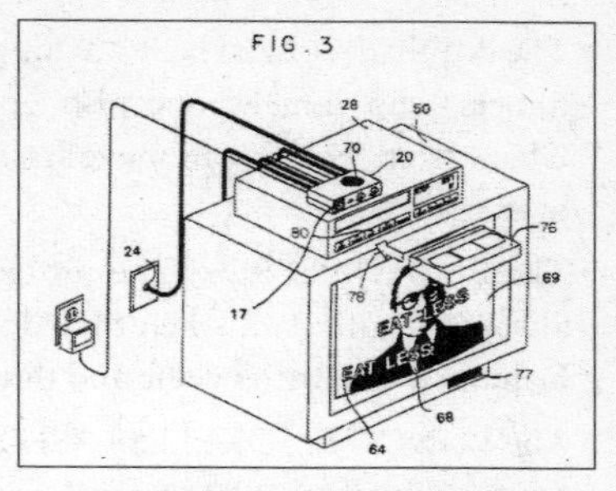

Ron Popeil has thirteen patents, including two for producing subliminal images on TV screens. This is a technical drawing from US Patent #5,221,962, "Subliminal Device Having Manual Adjustment of Perception Level of Subliminal Messages."

What were the inventions?

- The INSTANT-SHINE shoeshine spray.
- THE PLASTIC PLANT KIT, which consisted of "tubes of liquid plastic in a variety of leafy colors along with an assortment of metal plates with inverted leaf designs, stems, and green tape."
- DIAL-O-MATIC food slicer – "You slice through a tomato so thin, you can read a newspaper through it."
- The VEG-O-MATIC – "Slices and dices and juliennes to perfection. Slice a whole potato into uniform slices with one motion. Simply turn the ring and change from thin to thick slices. Like magic you can change from slicing to dicing. No one likes dicing onions. The Veg-O-Matic makes mounds of them fast. The only tears you'll shed will be tears of joy."

- The MINCE-O-MATIC – "With powerful vacuum grip!" Customers who bought one also got a "free" bonus: the Food Glamorizer, "which can make fresh lemon-peel twists fast like a bartender."
- The BUTTONEER – "The problem with buttons is that they always fall off. And when they do, don't sew them on the old-fashioned way with needle and thread. Use the Buttoneer!"
- The RONCO SMOKELESS ASHTRAY – When you put a cigarette in the ashtray, the smoke is sucked into a filtering system.
- MR MICROPHONE – A simple wireless microphone that could broadcast to any nearby FM radio. "It's practical and great fun for the whole family, and it's only $14.88. Buy two or three, they make really great gifts!"
- The INSIDE THE OUTSIDE WINDOW WASHER – This product didn't sell too well.
- The TRIM-COMB – A small plastic comb with a razor blade inside of it. "Now anyone can trim hair and eliminate costly haircuts. It trims, thins, shapes, blends, and tapers. All you do is comb."
- The RONCO BOTTLE AND JAR CUTTER – "An exciting new way to recycle throwaway bottles and jars into decorative glassware, centerpieces, thousands of things ... A hobby for Dad, craft for the kids, a great gift for Mom. The Ronco Bottle and Jar Cutter. Only $7.77."
- The POPEIL POCKET FISHERMAN – "Want to make a boy happy? Give him the Pocket Fisherman."
- The RONCO 5-TRAY ELECTRIC FOOD DEHYDRATOR – "A device for producing beef jerky, banana chips, soup mix, and even potpourri at home."
- The HULA HOE – "The weeder with a wiggle."
- CELLUTROL – "The beauty aid for buttocks, hips, and thighs."

- The GLH (Great Looking Hair!) FORMULA NUMBER 9 HAIR SYSTEM – Spray-on paint that covers bald spots.

In 1991 and 1993, Popeil obtained patents (US #5,017,143 and US #5,221,962) for new ways to produce subliminal advertising messages on TV screens.

In his autobiography Ron Popeil described the mystique of the inventor:

"We're celebrities, folk heroes, the common man (or woman) who has made good. Being an inventor (even if you're really [just] an innovator) seems to give you credibility. What's even better is if you can invent, innovate, and market. The combination of the three is sure to make you a media celebrity."

The autobiography begins with a simple thought that characterizes Popeil's spirit and his many fine inventions:

"I pushed. I yelled. I hawked."

And it worked. "I was stuffing money into my pockets, more money than I had ever seen in my life."

For his years of ceaseless invention, Ron Popeil won the 1993 Ig Nobel Prize in the field of Consumer Engineering.

The winner could not, or would not, attend the Ig Nobel Prize Ceremony. He carried on with his life's work of inventing and selling.

The Kitty and the Keyboard

"When a cat first places its paw down, the cat's weight plus the momentum of the cat's movement exerts pounds of force on the keyboard, primarily through the cat's paw pads. The cat's paw angles and toe positions also undergo complex changes while the paw lands on the keyboard. This forces keys and often key combinations down in a distinctive style of typing which includes unusual timing patterns. Cats' patterns of overall movement in walking or lying down also help make their typing more recognizable."

–from a technical description supplied by the manufacturer

THE OFFICIAL CITATION

THE IG NOBEL COMPUTER SCIENCE PRIZE WAS AWARDED TO

Chris Niswander of Tucson, Arizona, for inventing PawSense, software that detects when a cat is walking across your computer keyboard.

PawSense is available from BitBoost Systems, 421 E. Drachman, Tucson, AZ 85705, USA. (http://www.bitboost.com)

Chris Niswander (pronounced "nice-wander") is a computer scientist who is also editor of the Tucson Mensa Society newsletter. He took an intellectual approach to the fundamental cat/computer problem. First he stated the problem:

"When cats walk or climb on your keyboard, they can enter random commands and data, damage your files, and even crash your computer. This can happen whether you are near the computer or have suddenly been called away from it."

Having stated the problem, he then solved it. In Mr Niswander's words:

"PawSense is a software utility that helps protect your computer from cats. It quickly detects and blocks cat typing, and also helps train your cat to stay off the computer keyboard."

What people always want to know, after they ask "Why?," is "How does it work?"

Mr Niswander always answers the first question politely. Then he explains that "PawSense detects cat typing by weighing a combination of factors to achieve maximal speed and reliability. It analyzes keypress timings and combinations to distinguish cat typing from human typing. PawSense normally recognizes a cat on the keyboard within one or two pawsteps."

When PawSense detects a cat on the keyboard, it takes action, unleashing a blast of loud harmonica music. Alternatively, it unleashes a loud recording of Mr Niswander hissing, or one of several other sounds which some humans may find delightful, but most cats will not.

Mr Niswander says that the sounds are not effective against deaf cats, but that once a cat has been recognized, PawSense blocks the cat's keyboard input. In that event, it puts up a giant message on the computer screen: "CAT-LIKE TYPING DETECTED" and requests that you, or the cat, type the word "human." An illiterate cat might beat the system through a lucky combination of paw blows, but its odds of doing so are low.

Mr Niswander has submitted a patent application for PawSense. He says he plans a second product called "Baby-Sense," but that it requires an indeterminate amount of research and development, and so he is reluctant to name a date on which it will be available. For the meantime, he offers this advice to PawSense customers who want at least partial protection against their little progeny:

"If your baby bangs away with outstretched or open hands, or with fists, that makes keypress patterns close enough to those of cat paws, so PawSense should work relatively well. If your baby carefully pecks at only one key at a time, PawSense will recognize that your baby is indeed a human."

For protecting computers from feline catastrophe, and, as a bonus, giving minimal protection against babies, Chris Niswander won the 2000 Ig Nobel Prize in the field of Computer Science.

Mr Niswander traveled, at his own expense, from Tucson, Arizona, to the Ig Nobel Prize Ceremony. In accepting the Prize, he said:

"I'd like to thank my sister's cat Phobos for persuading me that this was a really, really good idea. I guess that's all I have to say. Thanks, Phobos, for persuading me that this was a really good idea."

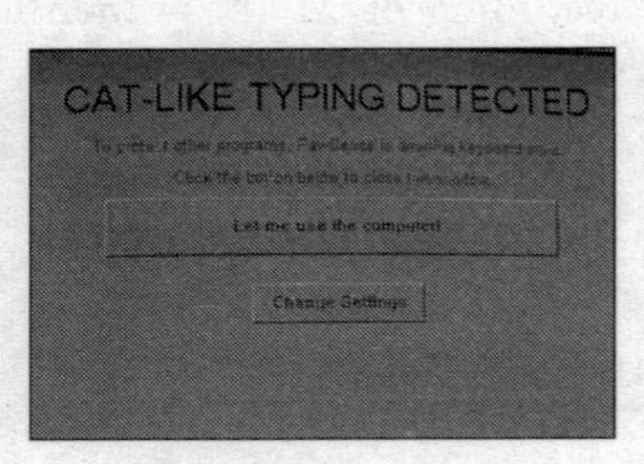

PawSense indicates that a cat is at work on the computer keyboard.

When Mr Niswander had finished speaking, Mr Leonid Hambro took elegant possession of the stage to deliver a personal tribute. Mr Hambro is the former principal pianist for the New York Philharmonic Orchestra, and for ten years after that he was the touring partner of pianist/comedian Victor Borge. The number he chose to play for this special occasion: Zez Confrey's 1921 composition *Kitten on the Keys*.

AutoVision

"Experiments suggest that primates derive pleasure from observing complex situations involving color, brightness, and movement. Rhesus monkeys, when given the means, will create movies in preference to still photographs. Modern psychobiological research is showing that a human's response is strongly affected by and may even be dependent upon movement in the environment. This movement creates the novelty and stimulation, implying the greater information content, which primates seem to desire. To keep a vehicle operator stimulated to the point of maximal responsiveness, then, it is advantageous to present a moving visual image in the general direction requiring the greatest attention."

- from US Patent #4,742,389, which describes an "Apparatus for promoting the vigilance of a motor vehicle operator"

THE OFFICIAL CITATION

THE IG NOBEL VISIONARY TECHNOLOGY PRIZE WAS PRESENTED JOINTLY TO

Jay Schiffman of Farmington Hills, Michigan, crack inventor of AutoVision, an image-projection device that makes it possible to drive a car and watch television at the same time, and to the Michigan state legislature, for making it legal to do so.

Jay Schiffman was granted five patents for his AutoVision technology: US Patents #4,742,389, #4,876,594, #4,884,135, and #4,937,665, all for an "Apparatus for Promoting the Vigilance of a Motor Vehicle Operator;" and US patent #5,061,996 for a "Ground Vehicle Head Up Display for Passenger"

Jay Schiffman realized before anyone else that roads would be much safer if drivers stopped trying to concentrate on their driving – instead, drivers should be watching television programs while they drive. Schiffman's simple invention makes that a snap.

Called "AutoVision," the invention can be installed in almost any automobile. An article in *Omni* magazine described it this way:

"It uses a projector, mounted near the dome light, which beams TV images to a matchbook-sized mirror lens near the windshield. Owing to an optical illusion, the picture appears to float above the horizon, a dozen feet in front of the car. It's like looking at a 12-inch TV set from across a room, except there's no TV and no room."

"Why does everyone consider this crazy?" Schiffman asked the reporter from *Omni*. "Is there some scientific basis for that belief? Absolutely not, if it's done with our patented methodology and configuration."

Schiffman was granted five patents on the technology.

US Patent #4,884,135 explains that "The invention overcomes 'road hypnosis' or other inattentiveness by maintaining the operator's interest, thereby putting the operator's psychological perceptive apparatus in a high state of readiness."

The patents, taken together, explain Schiffman's inventive reasoning. There are three steps to this reasoning.

The first logical step has to do with attention:

"Driving a motor vehicle is a task which demands relatively more visual than auditory attention. This explains why a person can successfully drive an automobile and simultaneously listen to a radio program. Any events that divert a driver's visual attention from the outside surroundings ahead of the vehicle immediately increase the probability of the vehicle being involved in a mishap."

The second logical step has to do with alertness:

"Since the time that they were first used in automobiles, radio receivers have served the dual purposes of providing entertainment and helping to maintain driver alertness.

These purposes are served on long-distance trips as well as shorter commuter trips, where the tedium of congested traffic conditions can lead to driver inattentiveness. The recent introduction of audio cassette players to motor vehicles has given the driver an additional method of entertainment and maintaining alertness. It is quite clear that a significant degree of driver attention may be sustained solely through the human auditory system."

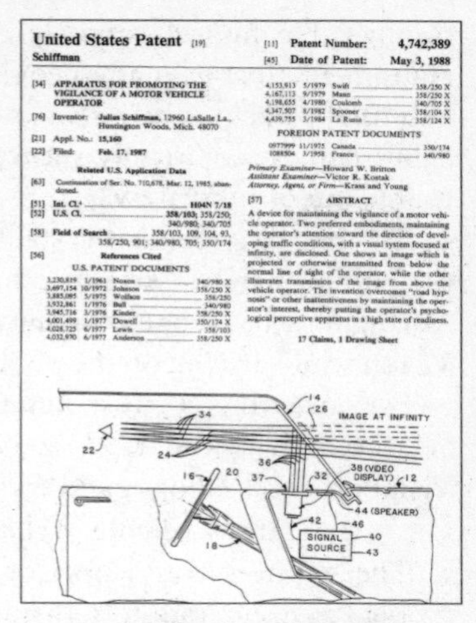

United States Patent [19]
Schiffman

[11] **Patent Number: 4,742,389**
[45] **Date of Patent: May 3, 1988**

[54] **APPARATUS FOR PROMOTING THE VIGILANCE OF A MOTOR VEHICLE OPERATOR**

[76] Inventor: **Julius Schiffman,** 12960 LaSalle La., Huntington Woods, Mich. 48070

[21] Appl. No.: **15,160**

[22] Filed: **Feb. 17, 1987**

Related U.S. Application Data

[63] Continuation of Ser. No. 710,678, Mar. 12, 1985, abandoned.

[51] **Int. Cl.⁴** **H04N 7/18**
[52] **U.S. Cl.** **358/103;** 358/250; 340/980; 340/705
[58] **Field of Search** 358/103, 109, 104, 93, 358/250, 901; 340/980, 705; 350/174

[56] **References Cited**

U.S. PATENT DOCUMENTS

3,230,819 1/1961 Noxon .... 340/980 X
3,697,154 10/1972 Johnson .... 358/250 X
3,885,095 5/1975 Wolfson 358/250
3,932,861 1/1976 Bull 340/980
3,945,716 3/1976 Kinder .... 358/250 X
4,001,499 1/1977 Dowell .... 350/174 X
4,028,725 6/1977 Lewis 358/103
4,032,970 6/1977 Anderson .... 358/250 X
4,153,913 5/1979 Swift .... 358/250 X
4,167,113 9/1979 Mann .... 358/250 X
4,198,655 4/1980 Coulomb .... 340/705 X
4,347,507 8/1982 Spooner .... 358/104 X
4,439,755 3/1984 La Russa .... 358/124 X

FOREIGN PATENT DOCUMENTS

0977999 11/1975 Canada 350/174
1088504 3/1958 France 340/980

Primary Examiner—Howard W. Britton
Assistant Examiner—Victor R. Kostak
Attorney, Agent, or Firm—Krass and Young

[57] **ABSTRACT**

A device for maintaining the vigilance of a motor vehicle operator. Two preferred embodiments, maintaining the operator's attention toward the direction of developing traffic conditions, with a visual system focused at infinity, are disclosed. One shows an image which is projected or otherwise transmitted from below the normal line of sight of the operator, while the other illustrates transmission of the image from above the vehicle operator. The invention overcomes "road hypnosis" or other inattentiveness by maintaining the operator's interest, thereby putting the operator's psychological perceptive apparatus in a high state of readiness.

17 Claims, 1 Drawing Sheet

Jay Schiffman's earliest patent for AutoVision.

The third, and conclusive, step in the logic is about what matters most – safety:

"The apparatus for maintaining vigilance by attracting an operator's visual attention ahead of the vehicle with a head-up visual display and entertainment system whose images are focused at infinity will promote highway safety."

The patent also says that "aggressive people may be influenced to drive more safely through its use."

Once he had designed and patented AutoVision, Jay Schiffman still had to build and demonstrate a working model. The goal was to persuade the major auto makers to include AutoVision in every car and educate their customers on the importance of watching television while driving.

Schiffman built a working model and installed it in his own car. He enjoyed taking people on demonstration outings. He himself quickly adapted to his new safety equipment. Just as many people, when starting their cars, habitually buckle their seat belts and turn on the radio, Schiffman automatically turned on the TV projector and roared out of his driveway.

The law enforcement system is designed to help ensure the public's safety. Certainly, that is the case in the state of Michigan, which enacted a law making it legal to use Auto-Vision while driving on the public roads.

For inventing a most entertaining safety device, Jay Schiffman won the 1993 Ig Nobel Prize in the field of Visionary Technology; and for making it legal to drive so safely, the state of Michigan shared in the honor.

The winners could not, or would not, attend the Ig Nobel Prize Ceremony. In turning down the invitation, Jay Schiffman said, "I don't see how this could help me or my company."

Schiffman was puzzled that people questioned the worth of his invention. "This isn't like cold fusion," he told a reporter, "I can demonstrate it. Even with a pornographic videotape, you can drive in traffic, no problem."

Schiffman's product never caught on with any of the major automobile manufacturers. Even the product name lost its significance. In 1998, Johnson Controls began marketing an automobile video display system called "AutoVision," but it is nothing like Jay Schiffman's invention. It is not a safety device, and it is not even meant for the driver. Advertisements say that "AutoVision enables rear-seat passengers to watch television, play video games, or surf the Internet."

And so a fantastic piece of technology goes ignored. The passengers may be entertained now, but the driving itself is far less entertaining than it would be if Jay Schiffman had his way.

Patenting the Wheel

"This invention relates to a device for facilitating transport of goods and persons. In particular, the device relates to a circular object which enables such goods and persons to be held above a surface and simultaneously moved with respect to the surface approximately parallel thereto."

– from Australia Innovation Patent #2001100012

THE OFFICIAL CITATION

THE IG NOBEL TECHNOLOGY PRIZE WAS AWARDED JOINTLY TO

John Keogh of Hawthorn, Victoria, Australia, for patenting the wheel in the year 2001, and to the Australian Patent Office for granting him Innovation Patent #2001100012.

"Melbourne Man Patents the Wheel," screamed the headline in the July 2, 2001 issue of the Australian newspaper *The Age.* The article told how this came about:

"A Melbourne man has patented the wheel. Freelance patent attorney John Keogh was issued with an Innovation Patent for a 'circular transportation facilitation device' within days of the new patent system being invoked in May.

"But he has no immediate plans to patent fire, crop rotation or other fundamental advances in civilization. Mr Keogh said he patented the wheel to prove the innovation patent system was flawed because it did not need to be examined by the patent office, IP Australia.

"'The patent office would be required to issue a patent for anything,' he said. 'All they're doing is putting a rubber stamp on it. The impetus came from the Federal Government. Their constituents claimed the cost of obtaining a patent was too high so the government decided to find a way to issue a patent more easily.'"

The Patent Office provides the two types of patents granted in Australia:

• A STANDARD PATENT gives long-term protection and control over an invention for up to 20 years.

• An INNOVATION PATENT is a relatively fast, inexpensive protection option available from IP Australia, and is the most recent in a range of other intellectual property rights. Protection lasts for a maximum of eight years.

The Patent Office states that "The estimated cost of an Australian standard patent including attorney fees is about $5,000 to $8,000." An innovation patent costs $180.

Mr Keogh obtained an innovation patent. Specifically, he obtained Innovation Patent #2001100012. Officially, his invention is called "a circular transportation facilitation device."

Commissioner of Patents Vivienne Thom has been quoted as saying, "to obtain the patent the applicant must make a declaration that they are the inventor. Obtaining a patent for a wheel would require a false claim, which is a very serious matter and would certainly invalidate the patent as well as amount to a misrepresentation on the part of the applicant and unprofessional conduct by any professional adviser."

The Patent Office's web site (www.IPAustralia.gov.au) advises patent applicants to examine the records of patents already on file. "Don't reinvent the wheel," says the web page. "Searching worldwide patent information can help you avoid wasting time and money duplicating work done elsewhere."

For being the first to patent the wheel in the 21st century, John Keogh won the 2001 Ig Nobel Prize in the field of

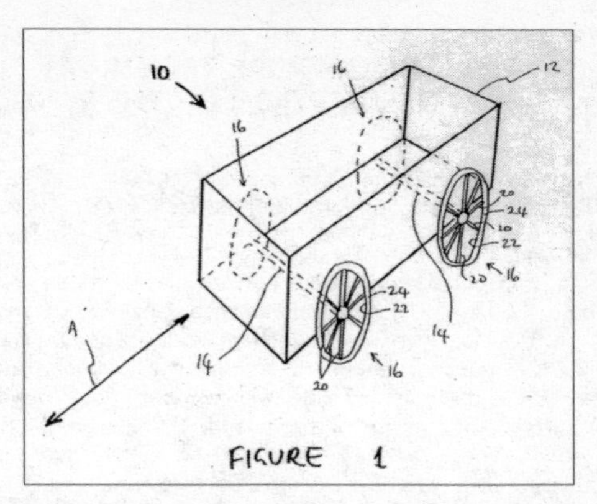

Figure 1 A technical drawing from the patent: "A perspective drawing of a cart incorporating a series of circular transportation facilitation devices in accordance with a preferred aspect of the present invention."

Technology; and for permitting him to do it, the Australian Patent Office shared the honor.

The winners could not, or would not, attend the Ig Nobel Prize Ceremony, but John Keogh did prepare and send a videotaped acceptance speech. In it, he said:

"When I sat down to write the patent specification for the wheel, I had one objective in mind: to expose a key weakness in Australia's new innovation patent system, which requires the Australian Patent Office to grant a patent for virtually anything that is applied for. The winning of an Ig Nobel Prize was not an outcome I foresaw. Obtaining a patent for the wheel has had some positive outcomes. The grant of the patent attracted publicity, both within Australia and around the world, highlighting a significant issue in our intellectual property laws. I can only hope that the awarding of the Prize contributes to the push to amend Australian patent laws to ensure the wheel will not again be patented."

IG NOBEL PRIZES

THE OFFICIAL TECHNICAL DESCRIPTION OF THE WHEEL

This is the technical description – now the official technical description – of a "circular transportation facilitation device":

BACKGROUND OF THE INVENTION

In the past, transportation of goods and services has been conducted in a number of ways. The predominant means has been transport of persons on foot, and carrying thereby of goods requiring transport. Other means of transport have included, in colder climates, skis, sleds, toboggans and the like, which slide over a smooth (low coefficient of friction) surface such as ice or snow, thus transporting the person and/or goods. These modes of transport have the advantage that when travelling down a sloped surface, free movement, that is, unassisted forward motion, is possible. The user is only required to apply effort to cause movement when travelling uphill or on a substantially flat plane, and this reduced effort helps the user move to the desired destination more quickly and more easily.

Unfortunately, such smooth surfaces for sliding over are not generally available in warmer climates where ice and snow do not form naturally. As such, and in the absence of alternatives, foot transport may be required. It would be useful if a device was available which enabled such unassisted forward motion on downhill slopes on surfaces having a much higher coefficient of friction than snow or ice.

SUMMARY OF THE INVENTION

In accordance with a first aspect of the present invention, there is provided a transportation facilitation device including:

• A circular rim;

• A bearing in which a hollow cylindrical member is rotatable about a rod within the hollow cylindrical member; and

• A series of connecting members connecting the circular rim with the hollow cylindrical member to maintain the circular rim and the hollow cylindrical member in substantially fixed relation; wherein

• The rod is positioned in an axis perpendicular to the plane of the circular rim, and substantially central of the circular rim.

PREFERRED ASPECTS OF THE INVENTION

In a preferred form of the present invention, a rubber layer is provided on an outer surface of the circular rim to allow smoother rolling of the circular rim over a surface on which it is placed, and to protect the outer surface of the circular rim. In a further preferred form of the invention, the rubber layer includes an inflatable tube.

Hellish Technicalities

In the race for Ig Nobel Prizes, there is one field in which the United States dominates, with all other countries struggling – and in many cases failing even to struggle – to compete. This chapter describes two Prize-winning achievements in the technical study of Hell.

- Who Is Going to Hell
- Mikhail Gorbachev is the Antichrist

Who Is Going to Hell

"Southern Baptist churches will be asked to kick off their new church year each October with a new emphasis on the reality of hell. Evangelist Bailey Smith announced June 14, in the annual sermon of the Southern Baptist Convention, the establishment of "Reality of Hell Sunday."

–report in the *Associated Baptist Press News*, June 16, 2000

THE OFFICIAL CITATION

THE IG NOBEL MATHEMATICS PRIZE WAS AWARDED TO

The Southern Baptist Church of Alabama, mathematical measurers of morality, for their county-by-county estimate of how many Alabama citizens will go to Hell if they don't repent.

The "Evangelistic Index" was published by the Southern Baptist Convention's Home Mission Board for its internal use. The report as a whole has not been disseminated to the public, but a significant portion was published in the September 5, 1993 issue of the Birmingham [Alabama] News.

The Southern Baptist Church of Alabama produced the first regional estimates of how many Alabamans are going to Hell. They based these on modern data-gathering and statistical methods. But the Church did not limit its concern to Alabama. They also calculated how many people from other places are going to Hell.

The estimates were a practical tool, a guide for where to concentrate the Church's evangelical efforts and where not to bother.

Any well-run modern business does this. A company that sells insurance or cereal or automobiles likes to let its sales force know how many dependable customers are in each region, how many potential new customers, and also how many marginal prospects – people not worth wasting time on. With this information, the sales force can focus its efforts productively. So it is with the Southern Baptist Church of Alabama. Spokesperson Martin King told the *New York Times*:

"If we were selling snow tires, we'd want to ask ourselves, 'Where are the people who need snow tires?' It's kind of a crass analogy, but where are the people who need the Lord? That's where we need to go."

The Church assumes that, in a given neighborhood, nearly all the Southern Baptists are already saved (they also assume that, people being people, a certain small percentage are damned idiots). Other Baptist and evangelical denominations are a mixed lot – some are still savable, others have irrevocably blown it. Most, but not all, Catholics are a lost cause. Non-Christians – Jews, Muslims, Hindus, Confucians, atheists, and others who refuse to accept Jesus – can be written off, evangelically speaking.

The Southern Baptist Convention's Home Mission Board did all the work on this. They devised a secret mathematical formula, estimating what percentage of each religious group will go to hell: X% of Southern Baptists, Y% of Episcopalians, Z% of Catholics, and so on. The percentages are based on experience and instinct. The Home Mission Board puts great faith in these estimates.

It was easy to find out how many people of each faith live in each Alabama county. A group called the Glenmary Home Missioners Board, in Ohio, periodically publishes a massive county-by-county survey of the entire United States. The Southern Baptist Convention fed the 1990 survey numbers into their secret formula. The result: the "Evangelistic Index," the now-celebrated, county-by-county estimate of how many

Alabama residents are, as the professionals put it, "unsaved."

The Evangelistic Index was not meant to be celebrated. Like any sales estimate, it was prepared for the organization's internal use only. But someone gave parts of it to Greg Garrison, a reporter for the *Birmingham News*, and the newspaper published a page-one article that began with the news that:

"More than 1.86 million people in Alabama, 46.1% of the state's population, will be damned to Hell if they don't have a born-again experience professing Jesus Christ as their savior, according to a report by Southern Baptist researchers."

The *Birmingham News* mentioned in passing that this was just the half of it – or, to put it more accurately, this was just the one-fiftieth of it:

"The Southern Baptist researchers who compiled the Evangelistic Index did not publicize their estimates for how many Episcopalians, Presbyterians, Lutherans, Methodists, Catholics and others are saved or lost. 'They're not going to reveal the formula,' [spokesman] Steve Cloues said. 'It's just to give you a feel on a state-by-state basis, trying to show there are some areas where the need is more severe than others.'"

Yes, the Southern Baptist Church of Alabama made estimates for how many people are going to Hell in every county – not just in the state of Alabama – but in all 50 of the United States of America. The country-wide figures have never been made public, but anyone who needs to know can reconstruct the numbers pretty easily (see box overleaf). And once cracked, the un-heavenly secret formula can easily be applied to almost any region of any country on earth.

For mathematically estimating who's going to be hot and

IG NOBEL PRIZES

HOW MANY ALABAMA RESIDENTS ARE GOING TO HELL
– a County-by-County Estimate

(Based on 1990 data. Note: the original *Birmingham News* article contained a typographical error, wrongly indicating that the data were from a different year.)

COUNTY	% UNSAVED	COUNTY	% UNSAVED
Autauga	47.4	Houston	39.6
Baldwin	56.3	Jackson	55.0
Barbour	48.0	Jefferson	42.8
Blount	48.3	Lauderdale	49.2
Bullock	36.1	Lawrence	52.0
Butler	30.0	Lee	53.4
Calhoun	41.2	Limestone	55.5
Chambers	43.4	Lowndes	38.8
Cherokee	46.0	Macon	47.3
Chilton	40.0	Madison	55.2
Choctaw	35.4	Marengo	23.1
Clarke	35.1	Marion	48.7
Clay	30.4	Marshall	48.2
Cleburne	37.0	Mobile	50.1
Coffee	39.5	Monroe	36.5
Colbert	41.3	Montgomery	44.9
Conecuh	31.6	Morgan	44.4
Coosa	47.9	Perry	33.2
Covington	36.5	Pickens	35.6
Crenshaw	30.9	Pike	46.6
Cullman	38.2	Randolph	46.0
Dale	55.1	Russell	47.2
Dallas	47.0	Shelby	63.5
DeKalb	45.8	St. Clair	51.6
Elmore	45.7	Sumter	42.9
Escambia	45.8	Talladega	43.9
Etowah	34.7	Tallapoosa	41.5
Fayette	41.5	Tuscaloosa	51.6
Franklin	53.8	Walker	47.0
Geneva	38.6	Washington	34.3
Greene	34.8	Wilcox	42.8
Hale	39.4	Winston	44.6
Henry	35.6	STATE AVERAGE	46.1

IGNOBEL PRIZES

ARE YOU GOING TO HELL?

If you want to know your chance of going to Hell, here's how to find out.

If you live in Alabama, simply see the box "How Many Alabama Residents are Going to Hell", opposite.

If you live anywhere else, here's how to crack the Southern Baptist Church of Alabama's secret mathematical code, and calculate your chances of going to Hell.

To crack the code exactly takes a small amount of mathematical skill, but anyone who has a little patience and is adept with computer spreadsheets can come very close.

- FIRST, choose a handful of Alabama counties.
- SECOND, for each of those counties, look up (in the chart facing) the Southern Baptist Church's estimate of how many residents there were unsaved in 1990.
- THIRD, for each of those counties, get the 1990 list of how many county residents were adherents to each religious denomination. You can obtain this list from the Glenmary Research Center (www.glenmary.org) or from the American Religion Data Archive (www.thearda.com).
- FINALLY, using whatever method you prefer (simultaneous equations, or a computer spreadsheet, or good old seat-of-the-pants rough estimation), figure out the Unsaved-Souls Percentage (USP) for each denomination. When you've got the correct percentages, then this equation will add up properly:

((Episcopalians' USP) x (number of Episcopalians))
\+ ((Catholics' USP) x (number of Catholics))
\+ ((Jews' USP) x (number of Jews))
\+ ((Muslims' USP) x (number of Muslims))
\+ ...
\+ ((Agnostics' USP) x (number of Agnostics))
\+ ((Atheists' USP) x (number of atheists))

= Total number of unsaved souls
estimated by the Southern Baptist Church

Once you've cracked the secret formula, you can look up the religious breakdown of any county in the United States and apply the formula to see how many people there will go to Hell.

You can also do this for any other country or region. Just obtain the religious breakdown for citizens of that country or region (the web site www.adherents.com is a good first place to look for this). Then apply the formula to it.

Bonne chance!

(Note: This is an interesting and instructive exercise for any school mathematics class.)

Baptists count the lost

46% of Alabamians face damnation, report says

By Greg Garrison
News staff writer

More than 1.86 million people in Alabama, 46.1 percent of the state's population, will be damned to hell if they don't have a born-again experience professing Jesus Christ as their savior, according to a report by Southern Baptist researchers.

The Southern Baptist Convention's Home Mission Board recently released its Evangelistic Index study, an estimate of the "lost" with a county-by-county tally across Alabama of how many souls the Baptists regard as doomed if they do not get saved before they die.

The Baptists' religious census, done nationwide to help the denomination know where it should intensify outreach efforts, counts many Catholics, Jews, non-born-again mainline Protestants, sect members and the religiously unaffiliated among those needing salvation.

Jefferson County leads the state in the number of lost souls, and Shelby County has the highest percentage of potentially hell-bound citizens, according to the Southern Baptist report. Jefferson County has 278,786 lost souls, 42.8 percent of its population; Shelby County has 63,080 unsaved people, or 62.5 percent, of its population, according to Home Mission Board estimates.

The saved and

Southern Baptist estimates for Alabama counties

Below 30% unsaved
30% to 50% unsaved
50% or more unsaved

Source: Southern Baptist Home Mission Board

Clowes said he expected to be inundated with requests for the report since it was publicized in state Baptist circles in July. Fewer than a dozen people asked for it, he said.

This 1993 article in the *Birmingham News* gave the damned their first plain look at the statistics.

who's not, the Southern Baptist Church of Alabama won the 1994 Ig Nobel Prize in the field of Mathematics.

The winners could not, or would not, attend the Ig Nobel Prize Ceremony.

As a tribute to them, the Ig Nobel Board of Governors sent a representative to the little town of Hell, Norway, to interview the local citizenry and solicit their well wishes. The town's highest official, the railroad stationmaster, asked that his congratulations be expressed at the Ig Nobel Prize Ceremony. And so it was that Terje Korsnes, the Norwegian Consul to Boston, appeared at the Ig Nobel ceremony, saying:

"I was asked to come here tonight and accept custody of this prize on behalf of the people of Hell, Norway. We were delighted to learn that so many people in the great state of Alabama will go to Hell. We have a special place in Hell for all of you."

Mikhail Gorbachev is the Antichrist

"This will not be a pleasant book to read."

–the opening sentence of *Gorbachev! Has the Real Antichrist Come?*

THE OFFICIAL CITATION

THE IG NOBEL MATHMATICS PRIZE WAS AWARDED TO

Robert W. Faid of Greenville, South Carolina, farsighted and faithful seer of statistics, for calculating the exact odds (710,609,175, 188,282,000 to 1) that Mikhail Gorbachev is the Antichrist.

His study was published in the form of a book: *Gorbachev! Has the Real Antichrist Come?* Victory House, 1988.

In 1988, Robert W. Faid solved one of the oldest and most famous problems in mathematics. Yet almost no one noticed. Cracking the nut that was nearly two millennia old, Robert W. Faid calculated the identity of the Antichrist.

In the rarified world of mathematicians, certain problems become the focus of intense debate. The Four-Color Map Problem caused rabid fascination until Wolfgang Haken and Kenneth Appel devised a solution in 1976. Fermat's Last Theorem was all the rage until Andrew Wiles solved it in 1993.

Haken and Appel became instantly famous among mathematicians. Wiles became a worldwide celebrity, his face appearing everywhere in newspapers and on television.

But little public approbation came to Robert W. Faid.

The Antichrist problem has been on the books since about the year 90, when the *Book of John* was published. The book contains four occurrences of the word "Antichrist." John 2:18, for example, says: "Dear children, this is the last hour; and as you have heard that the Antichrist

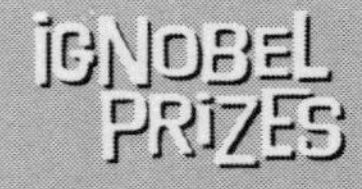

AN INVENTOR OF NOTE

Before he specialized in Antichrist arithmetic, Robert W. Faid was a full-time engineer. In 1976, he and a colleague were granted US patent #4,064,672 for an invention called a "post-applied waterstop connection." Here is the first page of the patent. The device's basic shape displays a harmony with the inventor's core values and outlook on life and engineering.

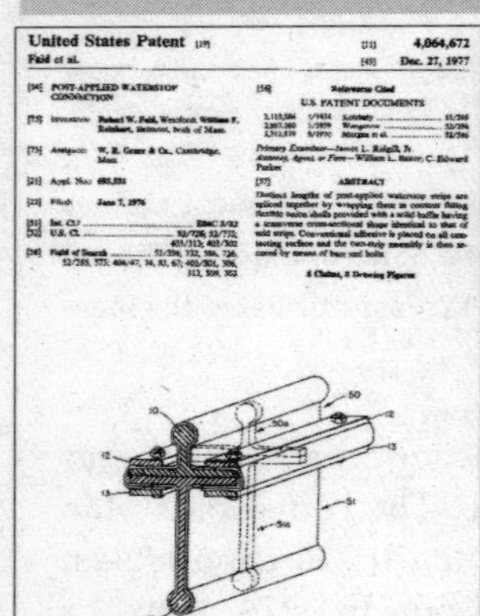

United States Patent [19] [11] **4,064,672**

Faid et al. [45] **Dec. 27, 1977**

[54] POST-APPLIED WATERSTOP CONNECTION

[75] Inventors: Robert W. Faid, [illegible]

[73] Assignee: W. R. Grace & Co., Cambridge, Mass.

[21] Appl. No.: [illegible]

[22] Filed: June 7, 1976

[51] Int. Cl. [illegible]

[52] U.S. Cl. [illegible]

[58] Field of Search [illegible]

[56] References Cited

U.S. PATENT DOCUMENTS

[illegible]

Primary Examiner—[illegible]

Attorney, Agent, or Firm—[illegible]

[57] ABSTRACT

[illegible]

[illegible]

Robert Faid's Prize-winning book, *Gorbachev! Has the Real Antichrist Come?*

is coming, even now many Antichrists have come. This is how we know it is the last hour."

Over the years, many amateur mathematicians joined the professionals in trying their hand at this delightful, yet maddeningly difficult, puzzle. Eventually it became a favorite old chestnut, something to be wondered at, but perhaps too difficult ever to yield up a solution.

In the 20th century, the problem rose to sudden popularity. In some circles it was now seen to be a fundamental problem of mathematics. Here and there, professional mathematicians hazarded solutions. But their solutions proved flawed. Then, as has happened so often in this rarified branch of science, an amateur stepped in and claimed the glory that had eluded the pros.

Robert W. Faid later wrote that the story "began about 1 a.m. on March 8, 1985, when I was awakened with a tremendous sense that something of great importance was about to happen." Working with almost demonic fury, he elucidated the factors necessary for a solution, reduced them to a set of 11 (or perhaps 22) numbers, and then multiplied everything together. Then glory, hallelujah, there it was. Robert W. Faid had solved the problem. He had calculated the identity of the Antichrist.

In retrospect, it seems almost absurdly simple: the Antichrist is Mikhail Gorbachev, with odds of 710,609,175,188,282,000 to 1.

How did Robert W. Faid do it? He knew that everyone would want to know, so he wrote a book explaining every first and last tittle and jot.

Robert W. Faid is a trained engineer. He is methodical. In the book *Gorbachev! Has the Real Antichrist Come?*, he carefully explains where each number comes from, and where it enters into the calculation. Then he summarizes all the factors. Here is the complete, concise list of the 11 (or perhaps 22) factors:

	Feature	*Probability*	*Odds*
1.	Mikhail S. Gorbachev in Russian = 666 x 2(+/-3)	95	94
2.	Mikhail S. Gorbachev in Russian = 46 x 29 (+-1)	15	14
3.	Mikhail Gorbachev in Russian = 46 x 27 (+-3)	6	5
4.	Mikhail S. Gorbachev in Greek equals 888 x 2 (+-1)	296	295
5.	Mikhail S. Gorbachev in Greek equals exactly 888	888	887
6.	Rise from obscurity over men of equal qualifications	2,000	1,999
7.	Soviet population exactly 276 million (Satan's number)	50	49
8.	Rules ten other kingdoms	10	9
9.	Exactly ten kings (Politburo members when elected)	10	9
10.	Exactly seven Warsaw Pact nations	10	9
11.	Being the eighth "king" or leader of the USSR	8	7

To the non-specialist – that is, to anyone without Robert W. Faid's education, experience and understanding, these numbers may be difficult to comprehend. For example, the difference between Robert W. Faid's "Probability" column and his "Odds" column is presumably rather subtle. But multiply the numbers together (see pages 206–8 of book *Gorbachev! Has the Real Antichrist Come?* for details), and out pops the final result: 710,609,175,188,282,000.

What does this number mean, this 710,609,175,188, 282,000? Robert W. Faid appreciates that many people are intimidated by statistics, and so he explains as simply as he can:

"The calculations indicate that the odds that Gorbachev is the actual and true Antichrist are: 710,609,175,188,282,000 to 1. That means that if you want to bet that Gorbachev is not the true Antichrist, you will be betting against odds of 710 quadrillion, 609 million, 175 billion, 188 million, 282,000.

"To get an idea just how large this number is, let us compare it with the population of the earth today. There are about five billion people living on the earth at this time. The mathematical probability of one person fitting all of the Antichrist prophesy and the hidden clues which we have examined that show that Mikhail S. Gorbachev indeed does fit, is the same as saying that only *one* person in 359,576,064 earths of the same population as ours would statistically meet them. If we were to assume, and correctly so, that the Antichrist would have to be an adult male, which is about one-fourth of the population, then this number of earths would be four times as many statistically, or 1,438,304,256 with the same population required."

Professional mathematicians find it difficult to argue with the logic of this.

Robert W. Faid's book was published in 1988. It did not receive the attention that its author had a right to expect. Moreover, the book did not come to the attention of certain high-level decision makers who might have made certain decisions differently, had they been aware of the knowledge contained therein. In particular, it probably was not given adequate consideration by the Norwegian Nobel Committee when that august body selected the winners of the 1990 Nobel Prizes.

And thus, it is not wholly inexplicable why in 1990 Mikhail Sergeyevich Gorbachev was awarded the Nobel Peace Prize.

Nor is it wholly inexplicable why in 1993 Robert W. Faid was awarded the Ig Nobel Prize in the field of Mathematics.

The winner could not, or would not, attend the Ig Nobel Prize Ceremony.

He continued his authorial career, producing *Lydia:*

Seller of Purple in 1991, *A Scientific Approach to Biblical Mysteries* in 1993, and *A Scientific Approach to More Biblical Mysteries* in 1995.

Jack Van Impe

Rexella Van Impe

IG NOBEL PRIZES

WHAT OTHER EXPERTS SAY

Jack and Rexella Van Impe, themselves Ig Nobel Prize winners (in 2001, for determining that black holes fulfill all of the technical requirements to be the location of Hell), produced a 90-minute video called *The Antichrist – Super Deceiver of the New World Order*, which, according to the promotional material, "answers some of the most intriguing questions of this or any other generation." Perhaps the most intriguing of those questions is this one: "What do Kaiser Wilhelm, Benito Mussolini, Adolf Hitler, Joseph Stalin, Nikita Krushchev, John F. Kennedy, Mikhail Gorbachev and Ronald Reagan have in common?" To obtain the answer, you must send $19.95 plus shipping/handling to Jack Van Impe Ministries.

Art & its Appreciation

Art need not be in the eye of a beholder. It can be removed to a box in a plastics factory; put in a room with pigeons; stored in the discarded, dirty, wet washrags of a troop of Scouts; imposed on a wheat field in the dark of night; or mounted on the business end of an animal. This chapter describes all those locations.

- The Birth of the Plastic Pink Flamngo
- Pigeons Prefer Picasso?
- A Good Deed: Trashing Art History
- Crop Circles
- Penises of the Animal Kingdom

The Birth of the Plastic Pink Flamingo

"I had to come up with something to keep from starving to death when I got out of art school."

–Don Featherstone

THE OFFICIAL CITATION

THE IG NOBEL ART PRIZE WAS AWARDED TO

Don Featherstone of Fitchburg, Massachusetts, for his ornamentally evolutionary invention, the plastic pink flamingo.

For the history of the plastic pink flamingo see the book *Pink Flamingos: Splendor on the Grass*, by Don Featherstone and Tom Herzing, Schiffer Publishing, 1999.

Don Featherstone changed the face of the planet. In 1957, he created the plastic pink flamingo, an object which, singly or in multitudes, has appeared on nearly every landscape on every continent on earth.

After graduating from art school, Don Featherstone's first and only job was at a local plastics factory. Union Products, in central Massachusetts, made two-dimensional lawn ornaments – dogs, frogs, ducks, and whatever else people were willing to buy. The new designer was asked to come up with three-dimensional versions of first a duck and then a flamingo. The ducks sold well. The flamingos made history. Their tackiness, their awkward ugly-non-duckling outer beauty, their gentle, glowing pinkness, their spindly metal shafts-for-legs, their low price, altogether somehow quietly but firmly caught the public's fancy and never let go.

People wonder about the thin, thin metal legs. Don Featherstone told an interviewer that "My original model had wooden dowels for legs, but they were too expensive to make and plastic wasn't strong enough, so we went with the metal rods. We once put out a model called the Flamingo

Don Featherstone, creator of the plastic pink flamingo, and his wife Nancy pose with and for admirers at the Eighth 1st Annual Ig Nobel Prize Ceremony. Nearly every year since Don won his Ig in 1996, the Featherstones have returned to the ceremony, to take a bow and mingle with their adoring admirers. Each year they have made and worn a new set of matching clothing, tailored to the theme of that year's ceremony. Their 1998 matching duct-tape tuxedo and evening gown, shown here, were much commented upon. Photo: Eric Workman/*Annals of Improbable Research*.

Deluxe. They looked very natural, with nice wooden yellow legs, but they wouldn't sell. It's almost like flamingo people think that the real birds have metal legs in their natural state."

As times changed, the plastic pink flamingo endured. That which had been 1950s beautiful became 1970s tacky, which was not a bad thing for the plastic pink flamingo business. Whereas the original generation of owners put flamingos on their lawns as a sign of cheery good taste, the next generation more often gave them as literally standing jokes. The Divinely and intensely offensive movie *Pink Flamingos* in 1972 somehow sanctified Don Featherstone's creation. Thereafter. the bird was a part of pop culture, forever old yet ever-new, always in the pink of good bad taste.

The man who made the flamingo always marveled at its appeal, convinced that dumb luck was the real secret to his success. A sweet and eternally bemused man, he married a bemused and sweet woman named Nancy. They decided to renew their marriage vows daily, not with words but with clothing. Every day since their wedding, Don and Nancy Featherstone have worn matching outfits, most of them designed by Nancy. The *New York Times* once featured a Valentine's Day fashion photo spread of the Featherstones modeling selections from their wardrobe.

Over the course of four decades, Don Featherstone designed more than 600 plastic products, a small universe of big plastic oddities and delights. None captured the public's imagination as fiercely as the flamingos. But perhaps nothing ever could. With that singular plastic mold, Don Featherstone achieved what only a handful of artists ever have: a firm lodging in the minds of men and women, and steady sales.

For going one better than Mother Nature, Don Featherstone won the 1996 Ig Nobel Prize in the field of Art.

Don and Nancy Featherstone came to the Ig Nobel Prize Ceremony wearing matching bright pink outfits. The audience greeted them with a rousing and thunderous ovation. The science community and the art community together were at long last paying heartfelt, joyous tribute to a man whose imagination had broken a mold, and whose hands had created one. The stage at Sanders Theatre was decorated with plastic pink flamingos by the dozen. The very moment the ceremony ended, hordes of audience members swooped down and carried off the birds, many pausing sheepishly to have their flamingos autographed by the creator.

Just after the turn of the century, Don Featherstone retired from Union Products. He and Nancy planned to do as the flamingo does, relax at home and from time to time appear in exotic places to bask in the sunlight.

What happened next was unexpected. Someone secretly monkeyed with the classic plastic pink flamingo.

Every Featherstone-designed flamingo manufactured since 1986 has Don Featherstone's signature on its butt. This feature was added on the flamingo's 30th birthday, immediately becoming a trusted symbol of authenticity. Several months after Don Featherstone's retirement, someone at Union Products quietly altered the flamingo mold to remove the signature.

When this became known, it was a shock to plastic pink flamingo lovers everywhere. Outraged connoisseurs, led by the *Annals of Improbable Research* and the Museum of Bad Art, called for a worldwide boycott of the altered birds. Reporters besieged Union Products with phone calls and visits – but the company drew a veil over its activities, refusing to answer phone calls. Such was the state of affairs as this book went to press. Consumers are advised that, before purchasing a plastic pink flamingo, they should examine its rump closely.

Pigeons Prefer Picasso?

"When we see paintings by Picasso and Monet, we can with some accuracy recognize which is Picasso's and which is Monet's, even if we have never seen the particular paintings before ... Can pigeons discriminate paintings of one artist from those of another artist?"

–from the research report "Pigeons' Discrimination of Paintings by Monet and Picasso"

THE OFFICIAL CITATION

THE IG NOBEL PSYCHOLOGY PRIZE WAS AWARDED TO

Shigeru Watanabe, Junko Sakamoto, and Masumi Wakita, of Keio University, for their success in training pigeons to discriminate between the paintings of Picasso and those of Monet.

Watanabe, Sakamoto, and Wakita's report is "Pigeons' Discrimination of Paintings by Monet and Picasso," *Journal of the Experimental Analysis of Behavior*, vol. 63, 1995, pp. 165–174.

While some people revile pigeons (see the section of this book about Ig Nobel Prize winner Lee Kuan Yew), others admire and study the birds' brainy behavior. A team of Japanese scientists demonstrated that pigeons can be taught to recognize recognized artists' art.

Presumably, some people learn on their own to appreciate the artworks of the great masters, but many learn from a teacher, either directly in a school or museum, or indirectly through books, magazines, and television programs. So is it with pigeons. Presumably, some birds learn on their own to appreciate the artworks of the great masters, but some gain their knowledge through formal instruction.

Shigeru Watanabe, a psychology professor at Keio University, and his colleagues Junko Sakamoto and Masumi Wakita set out to teach a group of pigeons how to discriminate paintings painted by Pablo Picasso (1881–1973) from those painted by Claude Monet (1840–1926).

The task was not an easy one. The pigeons had had no previous exposure to the works of either artist. As Watanabe, Sakamoto, and Wakita described it in their formal report, these were "eight experimentally naive pigeons."

The paintings used for teaching purposes included Monet's *Terrace at Saint-Adresse* (1866); *Poplars at Giverny* (1888); *Pond with Water Lilies* (1899); and *Il Palazzo Da Mula in Venezia* (1908); and seven others. The Picasso instruction consisted of viewing *Les Demoiselles d'Avignon* (1907); *Women Playing with a Ball on a Beach* (1932); *Nude Woman with a Comb* (1940); *Nude Woman Under a Pine Tree* (1959); and six others. The pigeons viewed each of these in the form of projected slides.

The birds' education continued with the viewing of videotapes. They watched videos of Monet's *River* (1868); *An Impression: Sunrise* (1872); *Station at St Lazare* (1877); and six others. They also viewed videos of ten of Picasso's paintings, ranging in style and content from *Donna con Ventaglio in Poltrona* (1908) to *Donna Dalle Mani Intrecciate* (1909) to *Dance* (1925).

The lessons were not unlike those presented in collegiate art history courses. The pigeons watched slides presented in random order, each slide being projected for 30 seconds, and a 5-second interval between slides. The whole time, a loudspeaker bathed them in 70 decibels of white noise.

Half the group were given hemp seeds every time they saw a Monet painting, but not when they saw a Picasso. The other half of the group were given seeds whenever they saw a Picasso, but nothing while a Monet was on display.

This regimen were repeated every day until the birds were able to achieve a score of 90% on a test, and do it two days in a row. The test consisted of being shown the paintings one more time and having to peck a key when they saw one painter's paintings, but keep their beaks shut when they saw anything else.

After demonstrating a basic mastery of the material in two to three weeks, the pigeons then had to take more advanced tests, first watching slides that were projected out of focus, then watching slides that were upside down. The former test was a piece of cake. In the latter test, the birds demonstrated that they could recognize Picasso paintings projected upside down, but they had trouble when this was tried with the Monets.

On balance, the results were about as good as any art teacher could reasonably hope for, for any group of students.

And so, for their perceptive work with pigeons, Picasso, and Matisse, Shigeru Watanabe, Junko Sakamoto, and Masumi Wakita won the 1995 Ig Nobel Prize in the field of Psychology.

The winners could not, or would not, attend the Ig Nobel Prize Ceremony.

As time went on, Shiguru Watanabe shifted his interest from pigeons to sparrows and from paintings to music. In 1999, he and a colleague issued a report explaining that they had taught seven sparrows to distinguish between the musical works of Johann Sebastian Bach (1685–1750) and those of Arnold Schoenberg (1874–1951). They also taught a different group of sparrows to distinguish between the music of Antonio Vivaldi (1678–1741) and Eliot Carter (1908–).

In 2001, Watanabe returned to the field of his previous triumph – the study of pigeons studying paintings. He doubly extended his earlier work, by including artistic styles beyond those of just Monet and Picasso, and by comparing the pigeons' ability with that of a different creature. In a report in the research journal *Animal Cognition*, Watanabe described the project this way:

"The author has previously reported that pigeons can discriminate paintings by different artists. Here, the author replicated the previous findings, carried out additional tests and compared discrimination by pigeons with that of four university students (aged 19–21 yrs). In Experiment 1, pigeons were trained to discriminate between paintings by Van Gogh and Chagall ... In Experiment 2, human subjects were tested with the same paintings ... [The results] suggest that the visual cognitive function of pigeons is comparable to that of humans."

A Good Deed: Trashing Art History

"M. Jean Clottes, inspecteur général de l'archéologie, nous a fait remarquer que les graffitis sont parfois tres anciens et ont alors une valeur de témoignage historique."

–from a report in *Le Monde*, March 24, 1992

THE OFFICIAL CITATION

THE IG NOBEL ARCHAEOLOGY PRIZE WAS AWARDED TO

Les Eclaireurs de France, the Protestant youth group whose name means "those who show the way," fresh-scrubbed removers of graffiti, for erasing the ancient paintings from the walls of the Mayrières Cave near the French village of Bruniquel.

The Boy Scouts and Girl Scouts of America, the Pfadfinderinnen und Pfadfinder, the Scouts Tunisiens, the Qatar Boy Scouts, the Guides et Scouts de Monaco, the Boy Scouts of Liberia, the Pfadfinder und Pfadfinderinnen Liechtensteins, the Escoteiros do Brasil, the Esploratori ed Esploratrici Italiani, the Bulgarskite Skauty, the Scouts of China, the Eclaireurs et Eclaireuses du Burkina Faso – in some 150 nations, boys and girls participate in scouting groups that work to do good deeds for their fellow citizens.

Most of these scouting groups strive to follow the motto, be it in English or a translation in their own language: "Be Prepared!"

In France in 1992, when their leaders asked les Eclaireurs de France to tackle a dirty job, les Eclaireurs were indeed prepared.

The troop leaders brought les Eclaireurs to visit la Grotte des Mayrières Supérieures, a cave in the Tarn-et-Garonne region in southern France. They directed the scouts to clean off the graffiti that covered the cave walls. Les Eclaireurs scrubbed off the graffiti.

What is graffiti to some is great art to others. In the case of this particular graffiti, it was great historical art.

The history is something the Scouts learned about only after they had finished their cleaning job. In 1952, a group of spelunkers made a fabulous historical discovery. The Grotte des Mayrières Supérieures has a long winding cavern, the walls of which are – or rather, were – adorned with a variety of ancient paintings. The most spectacular were not far from the entrance – two magnificent drawings of bison, one shown in a front-on view, the other in profile.

Archaeologists estimated the paintings to be somewhere between 10,000 and 15,000 years old. They were the only such paintings ever found in that part of France. Thanks to les Eclaireurs de France, they became the only such paintings ever lost there.

For their enthusiastic, and officially approved work, les Eclaireurs de France won the 1992 Ig Nobel Prize in the field of Archaeology.

The winners could not, or would not, attend the Ig Nobel Prize Ceremony.

Crop Circles

"The greatest of all physicists, Albert Einstein, proved that photons exist, but they do not exist in time ... He also stated that everything in existence is based upon the photon ... This would seem to support the theory that the circles are created by an unknown force field manipulated by an unknown intelligence."

– crop circle investigator Pat Delgado, in his book *Circular Evidence*

THE OFFICIAL CITATION

THE IG NOBEL PHYSICS PRIZE WAS AWARDED TO

David Chorley and Doug Bower, lions of low-energy physics, for their circular contributions to field theory based on the geometrical destruction of English crops.

There are many books about crop circle research. Here are two that make fairly opposite assessments: *The Secret History of Crop Circles: the True, Untold Story of the World's Greatest Mystery!* by Terry Wilson, The Centre for Crop Circle Studies, 1998.*Round in Circles: Poltergeists, Pranksters, and the Secret History of the Cropwatchers*, by Jim Schnabel, Prometheus Books, 1994.

They were mysterious, baffling, perhaps inexplicable. They arrived, always, in the dark of night, and they were in a way lovely – flattened circles of wheat, grass, or corn in the fields of England. Over the course of a decade, the circles cropped up in more and more places, and with all sorts of adornments to their shapes. The press, the populace, and the scientific community speculated like mad. What natural, unnatural, or exotically scientific forces could be bringing these wondrous things into being?

The answer, it turned out, was two guys, named Doug and Dave.

What They Did

As a lark one evening in the 1970s, Dave Chorley and Doug Bower created what history may regard, should it ever bother to think about such things, as the world's first official crop circle. Doug owned a picture-framing shop in Southampton. Dave was an amateur artist. The pair of them often drank together, and occasionally pulled off little pranks. One of their favorites was to go out after dark and make circles in the local fields, then see whether anyone noticed.

This particular little stunt was fun enough that they repeated it several times, until one day an article appeared in a local newspaper, the *Wiltshire Times*. That first news report described the strange circle ("the grain had been flattened, with the stalks all lying in a clockwise direction"), and quoted a local farmer as saying, "I have never seen marks like it before." The field with the odd doings was in an area with an appropriately odd name – Cheesefoot Head.

With the public now paying attention to their work, the boys went into overdrive. Chorley and Bower had used a metal bar to make the first crop circles. Before long, they made a great technical leap, switching to ropes and wooden planks, which were lighter and gave them longer reach.

As the boys churned out further masterpieces, the press fed a frenzy that spread throughout the country and then around the world. Crop circles somehow appeared here, there, and ultimately on every continent except Antarctica. One by one, the nations of the earth saw patterns appear in their fields. Scotland and Wales, Ireland and France, Holland, Sweden, Italy, Germany, Switzerland, Romania, Canada, the US, Mexico, Kenya, Argentina, Uruguay, New Zealand, Australia, Japan, the USSR.

Speculation abounded – what was causing these crop

circles? Helicopters, said some. Peculiar weather conditions, said others. Hitherto unknown scientific force fields, supposed more than a few. By far the largest group of crop circle enthusiasts believed or hoped that UFOs were involved. There were skeptics who shrugged it off as a hoax, but they were generally scoffed at by the weather, force-field, and UFO advocates.

The helicopter theorists soon dropped out of the running, because helicopters are a little too noticeable to have gone unnoticed.

The weather theorists published detailed reports in the *Journal of Meteorology* and elsewhere, explaining their belief that the crop circles were a predictable aerodynamic consequence of the actions of large central whirlwinds accompanied by as many as four satellite whirlwinds acting in precise

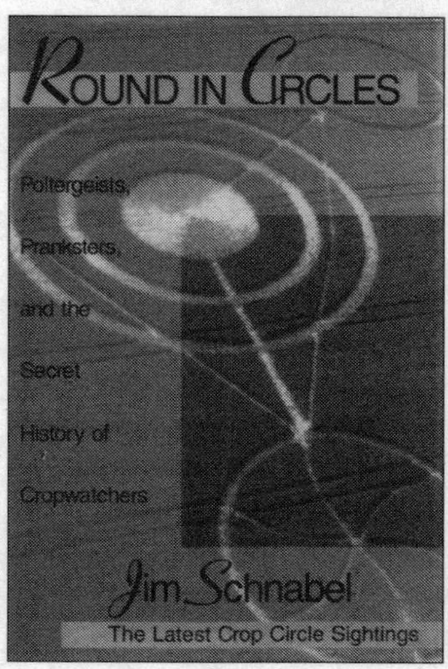

Books expressing opposite views about crop circles.

concert. The theorists had more complicated theories, too.

The unknown force-field advocates had a field day, so to speak, offering home-grown explanations of quasi-spherical plasma vortices, static charge inflows, radioactively-pitted wheat-stalk cell walls, and anomalous electromagnetic agitation.

As time went on, new crop circles appeared with shapes of increasing complexity, resembling animals, insects, keys, snowflakes, folk art, religious symbols, pinwheels, sprockets, and patterns from graduate mathematics textbooks. Chorley and Bower were responsible for some, but by no means all, of these. At one point they became so miffed by all the Doug-and-Dave-come-lately crop circle makers that they produced a giant wheat-field pattern spelling out the phrase "WE ARE NOT ALONE."

A three-issues-per-year journal called *The Cereologist* (which later changed its name to *The Cerealogis*t, and still later changed back to *The Cereologist*) was created to chronicle all this research. *The Circular*, *The Crop Watcher*, and other competing magazines sprang into existence. Books and television specials abounded. Crop circle organizations grew hither and yon.

By 1991, Chorley and Bower had been making crop circles long enough that, for them, the thrill was gone. They were also getting a little annoyed that a fair number of people were making money off crop circles – giving tours, selling souvenirs, writing books – but they themselves weren't. For whatever reasons, they phoned up reporters, told their stories, demonstrated their techniques, and became minor celebrities.

For all this, Dave Chorley and Doug Bower won the 1992 Ig Nobel Prize in the field of Physics.

The winners could not, or would not, attend the Ig Nobel Prize Ceremony.

After Chorley and Bower's efforts were made public, many crop circle enthusiasts became not-so-enthusiasts.

Some, though, dismissed Dave and Doug as boastful, interloping, anti-scientific troublemakers. Today, intrepid bands of crop circle investigators still continue to search for an explanation. Theirs is a tough row to hoe – and they know it. A book called *The Secret History of Crop Circles*, published by the Centre for Crop Circle Studies in 1998, makes the case against the Chorleys, the Bowers, and all those other cynics and skeptics:

"There is reason to believe that by the early 1990s a campaign was begun to destabilise crop circle research and convince the public of its unimportance."

Penises of the Animal Kingdom

"Penises of the Animal Kingdom is a comparative anatomy chart featuring the male copulatory organs of several animals, including man. The illustrations were rendered with close attention to proportion and scale, the sizes determined by the average physical dimensions of the genitalia of adult males. All organs are depicted erect at one-fifth actual size."

–from the descriptive insert accompanying the poster

THE OFFICIAL CITATION

THE IG NOBEL ART PRIZE WAS PRESENTED JOINTLY TO

Jim Knowlton, modern Renaissance man, for his classic anatomy poster Penises of the Animal Kingdom, and to the US National Endowment for the Arts for encouraging Mr Knowlton to extend his work in the form of a pop-up book.

The original *Penises of the Animal Kingdom* poster is now out of print. If and when Jim Knowlton completes work on the enhanced version, it will be available from him c/o: Scientific Novelty Co., P.O. Box 673, Bloomington, IN 47402, USA.

Jim Knowlton produced one of the most specialized, most innovative, most beloved, and most correct anatomy charts of the 20th century. It remains seared in the minds and hearts of everyone who has ever seen it.

In 1984, Jim Knowlton was a graduate student in physics at Columbia University. In a chance conversation, he heard that certain snakes have two penises, and that the cat penis is covered with sharp barbs. There ensued a detailed discussion about the oddities of animal penis morphology. By the end of the evening, Mr Knowlton had conceived the idea of a poster depicting the comparative anatomy of animal penises.

Penises of the Animal Kingdom is a comparative anatomy chart featuring the male copulatory organs of several animals, including man. The illustrations were rendered with close attention to proportion and scale, the sizes determined by the average physical dimensions of the genitalia of adult males. All organs are depicted erect at one-fifth actual size.

Each penis has certain outstanding features. The **human** organ possesses a well defined glans, or tip. This mushroom-shaped end is one of the most developed glandes of the animal kingdom.

The **dog** penis has a bulbous enlargement that is present only during erection. This bulb is the reason dogs "get stuck" while copulating. The female contracts her vagina around the trapped penis to extract seminal fluids.

Hyenas are well known for the similarity of the male and female genitalia. A female's erect clitoris is nearly identical, in both size and shape, to a male's penis. Covering the glans of each organ are sharp, backwardly directed spines.

The penises of the **goat**, **ram** and **giraffe** have extensions of the urethra. The urethras of the giraffe and ram can extend several centimeters beyond the glans of the penis, forming a pliant worm-like tube.

The **porpoise** has a remarkable penis. The copulatory part of the organ is jointed, allowing the tip to rotate or swivel. The animal has voluntary control over this action and uses the fingerlike appendage to manipulate and investigate objects in its environment.

Perhaps the oddest penis is that of the **pig**. During erection, the end of the penis convolutes into a corkscrew bearing an uncanny resemblance to the animal's coiled tail. The helical end of the erect organ conforms to the twisted contours of the female's vagina.

The **horse** penis is similar to that of the human; it also has a well defined glans. A dissimilar feature is a slight extension of the urethra.

The **bull** penis has an interesting history. Because of its rope-like consistency and proportions, it was used in the Middle Ages as a flogging stick. Today in some parts of the world it is dried and used as a walking cane.

The **elephant** has a very muscular penis. More than half of the curved organ forms the pendulous portion, yet only the very end penetrates the hard-to-reach vagina of the female during copulation.

Whales have the largest penises of all animals. A blue whale penis can measure thirteen feet in length and one foot in diameter. The poster depicts the sperm whale penis with a length of over seven feet.

To order: Send $9.95, plus $3 for postage and handling, to Scientific Novelty Co., Post Office Box 673, Bloomington, IN 47402. Please allow two weeks for delivery.

The "insert of descriptive text" that accompanied the original version of the classic anatomy chart *Penises of the Animal Kingdom.* It is reproduced here with the kind permission of Jim Knowlton.

He conducted intensive research at various university and public libraries in New York City, studying photographs, printed descriptions, and statistical data pertaining to the penises of numerous different species.

To convey the scientific basis and intent of the poster, Mr Knowlton chose to render the illustrations using a "pen-and-ink" line-drawing style reminiscent of the vintage reference work *Gray's Anatomy.* During the last week of December 1984, he arranged an initial printing of

1,000 copies, and then offered them for sale.

Penises of the Animal Kingdom was an instant hit. Mr Knowlton moved to Bloomington, Indiana, and founded a one-person company called Scientific Novelties, with the poster as its chief product.

Biology teachers, veterinarians, and penile pedagogues were immediately much taken with the poster. In an era when many students believed science to be a demanding and unstintingly rigid discipline, *Penises of the Animal Kingdom* offered a straightforward and interesting thrust.

The chart was wildly popular, but it did not meet with universal approval. In some medical schools, anatomy professors declined to adopt it for official use, perhaps because anatomy teachers must cover a tremendous amount and variety of material in a very short time, and thus are wary of visual aids which are overly specialized.

"In my opinion, though, the most interesting and unexpected response to the poster has been from the supposedly liberal media establishment," Knowlton lamented to interviewers. "Though a few mainstream magazines will carry advertisements for the poster, a vast majority of publications refuse the ads. Even self-proclaimed champions of tolerance and free expression, including *Playboy*, *Cosmopolitan*, and *The Atlantic*, have refused to advertise the chart."

In 1992, Mr Knowlton read about the National Endowment for the Arts (the NEA), a United States government agency devoted to encouraging and funding the arts. The NEA had recently provided money for Robert Mapplethorpe's anatomical photography and Andres Serrano's *Piss Christ*, a photograph of a crucifix suspended in urine. Knowlton hoped that they might be willing to fund an artistic project that was both more staid in style, and more scientific in content.

He telephoned the NEA's headquarters in Washington, DC, told them about the *Penises of the Animal Kingdom* poster, and explained that he would like extend that work in the form of a pop-up book. The NEA did not hang up on Mr Knowlton. They and he discussed his plan and his need, and Mr Knowlton received encouragement to apply formally for money.

And so, for their scholarly contributions to science and art, Jim Knowlton and the US National Endowment for the Arts shared the 1992 Ig Nobel Prize in the field of Art.

Mr Knowlton traveled from Bloomington, Indiana, to the Ig Nobel Prize Ceremony, at his own expense. His co-winner – the National Endowment for the Arts – could not, or would not, send a representative to attend the Ig Nobel Prize Ceremony. Here is the full text of Mr Knowlton's acceptance speech:

"In this century, a rigid barrier has been erected between art and science. As both an artist and a scientist, I believe this to be a dangerous thing.

"My seminal work is a comparative anatomy chart featuring the male copulatory organs of several animals from man to whale. I created the chart during my graduate studies at Columbia University. I have always been drawn to ambiguous and understated conceptual works. *Penises of the Animal Kingdom* exposes and exploits the tension between subject (the phallus) and dry context (the classical anatomy chart), stimulating artist and scientist alike to re-examine prevailing societal attitudes.

"It is my hope and dream to develop a pop-up book version of *Penises of the Animal Kingdom*. I have spoken at length with the National Endowment for the Arts. They have strongly encouraged me to apply to them for funding, and are advising me as to how they and I can best work together.

"On behalf of art, and on behalf of science, and on behalf of the members of the animal kingdom, I thank you."

At the conclusion of the ceremony, Mr Knowlton was besieged by animal enthusiasts, admiring women (aged 19–93), and anatomists, all seeking autographs, advice, and posters.

Two years later, at the insistence of adoring fans, he returned to the Ig Nobel Ceremony. In answer to the frenzied cheers of *Penises of the Animal Kingdom* devotees, he took the podium and said:

"The chart has a dual appeal. Though it is scientific in conception and appearance, it obviously has a humorous side as well. In creating the chart, I chose to present a provocative subject – the phallus – in the clinical and dry context of an anatomy chart. Though the tension between subject and context is a source of the chart's humor, I hope that in the end this conflict is resolved through a re-examination of the prevailing societal attitudes that uphold the phallic mystique, and that the poster contributes to a better appreciation of the true biological significance of the penis."

In 1996, he returned to Cambridge, to help honor the new crop of winners at the Sixth 1st Annual Ig Nobel Prize Ceremony. He used the forum to discuss in public the concept of biodiversity. "My research," he told the Nobel Laureates and the 1,200 other distinguished guests assembled in Sanders Theatre, "has revealed a surprising diversity in the members of the animal kingdom. Some of the members are large, others are less so. And while I'm here at

Harvard, I'd like to put in a plug for one of my favorite books – Professor Stephen Jay Gould's *The Mismeasure of Man.*"

Jim Knowlton sold more than 25,000 copies of the original poster before retiring, temporarily, from the worlds of art and anatomy. In 2001, he commenced work on a new, enhanced version of *Penises of the Animal Kingdom.*

Fragrances – The Good, The Bad, The Ugly

Smell is believed to be the most primitive of our senses. This chapter describes three attempts at sophistication:

The good:
- The Self-Perfuming Business Suit

The bad:
- Filter-Equipped Underwear

The ugly:
- A Man Who Pricked His Finger and Smelled Putrid for 5 Years

THE GOOD: The Self-Perfuming Business Suit

"SEOUL, South Korea – As Mr Lee Soo-bum nears home after an evening out with the guys, he shimmies, shakes and occasionally rubs his chest. Then at his apartment door, the 39-year-old film company executive sniffs, smiles in satisfaction, and greets his wife. Although he's been drinking with colleagues in a smoke-filled bar, Lee doesn't reek of booze and cigarettes. In fact, he smells downright sweet. 'This new suit helps keep peace at home,' Lee says of his fashionable beige wool suit. It smells like lavender – and the more he moves, the stronger the scent becomes."

–from a 1998 Associated Press report

THE OFFICIAL CITATION

THE IG NOBEL ENVIRONMENTAL PROTECTION PRIZE WAS AWARDED TO

Hyuk-ho Kwon of Kolon Company of Seoul, Korea, for inventing the self-perfuming business suit.

When businessmen come home to their wives after a hard night of drinking and smoking for professional purposes, they can, through no fault of their own, look and smell bad. But they don't have to. Hyuk-ho Kwon has made it easy for them to look good and smell good, no matter how late they stay out. Mr Kwon is the inventor of the self-perfuming business suit.

Hyuk-ho Kwon is an accomplished, yet quietly personable, employee of the Kolon Company. In all of the organization's 21 subsidiaries, which include enterprises ranging from textiles and chemicals to construction, trading, financial services, information processing and communications, Mr Kwon is perhaps the only individual who could have perfected this peculiarly stylish technology.

Nobel Laureate Dudley Herschbach encourages an audience member to rub the sleeve of his self-scented business suit. Other self-perfuming-suit-wearing Laureates are visible behind him. Photo: Jenny Lolley/*Annals of Improbable Research*.

The suits come in pine, lavender, and peppermint. They are impeccably tailored of very fine-quality wool.

The fabric is soaked in microencapsulated scent, and therein lies the delightful rub. Mr Kwon recommends vigorously rubbing the sleeve whenever a new burst of freshness is wanted. But the wearer seldom needs to make an effort. The scent is always discernable – and especially so whenever the man walks, or moves in any way, because his every slight motion breaks open a few more of the many millions of microscopic scent capsules.

The suits are made to keep their character through 20 or more dry cleanings – an estimated two to three years of typical use.

Kolon has two major competitors in the self-perfuming-business-suit market – L.G. Fashion, and Essess Heartist. All of them are based in Korea. The Korean fashion industry has transformed itself from a staid manufacturer of foreign designs into an aggressive innovator. When the self-perfuming-business-suit market has established itself throughout the rest of Asia and on the other continents, the newest of the new will still, fashion analysts expect, be coming from Seoul.

Even in its earliest days, the self-perfuming-business-suit market was not limited to married executives. Single, young executives-of-the-future also recognized the sweet smell of success.

"My lavender [scented] suit helps me keep the peace at home," office worker Lee Gyung-wook told Reuters news agency, echoing countless of his married, self-perfuming-business-suited superiors. "Without it, my parents would be all over me because of the stench of soju [Korean liquor] and spicy side dishes after nights out with my colleagues

Hyuk-ho Kwon, inventor of the self-perfuming business suit, learns from Miss Sweetie Poo that his speech is of sufficient duration. Photo: David Holzman/*Annals of Improbable Research*.

and friends," Lee said. "It's a huge relief since I no longer have to pour cheap cologne all over me. All I have to do now is just shake and shimmy in front of my house, and then go in with a frown on my face, saying, 'Man, I hate night shifts.'" Reuters also interviewed Moon Chol-ho, 28, who said: "After a hard day's work, we don't smell good with sweat. It's nice to wear the scented suit. It doesn't give off an unpleasant smell to others."

For engineering fail-safe fragrances into the traditional business suit, Hyuk-ho Kwon won the 1999 Ig Nobel Prize in the field of Environmental Protection.

Mr Kwon traveled from Seoul, Korea, at his company's expense, to attend the Ig Nobel Prize Ceremony. The Kolon Company generously made self-perfuming business suits for the five Nobel Laureates who participated in the ceremony, and for the master of ceremonies. Self-scented business suits dominated Sanders Theatre. In accepting his Ig Nobel Prize, Mr Kwon said:

"Thank you. The stronger you rub it, the stronger it smells. Thank you. It's my greatest pleasure to have the honor of receiving this prize. I just put my faith in God, hoping that there will be fragrance in my life. However, I came to realize that fragrance is in my suit. So, since I expect my lifetime to be fragrant, I hope every one of you may have fragrance in your life."

THE BAD: Filter-Equipped Underwear

"This invention relates to protective underwear, specifically to the filtering of foul smelling human flatus."

–from US Patent # 5,593,398

THE OFFICIAL CITATION

THE IG NOBEL BIOLOGY PRIZE WAS AWARDED TO

Buck Weimer of Pueblo, Colorado, for inventing Under-Ease, airtight underwear with a replaceable charcoal filter that removes bad-smelling gases before they escape.

Under-Ease is described in US Patent # 5,593,398 ("Protective Underwear With Malodorous Flatus Filter"). The product is available in men's and women's models from Under-Tec Corporation of Pueblo, Colorado (telephone 888-433-5913 in the US, 719-584-7782 outside the US), www.under-tec.com.

After years of olfactory suffering, Buck Weimer chose to invent a solution to his wife's intermittent, explosive problem. Unembarrassed, the couple were then kind enough to share that solution with the world.

A June, 2001, report in the *Denver* [Colorado] *Post* told the story clearly:

"Buck Weimer, 62, of Pueblo, [tells] what happened after a huge Thanksgiving dinner more than six years ago. He and his wife, Arlene, 57, who suffers from Crohn's disease, a form of inflammatory bowel syndrome, were lying under the covers when she let go a bomb.

"'I'm laying in bed with her, sort of suffering silently,' he said. 'Out of the silence came determination. Something had to be done.' More than six years later, Buck Weimer has a new invention: Under-Ease, airtight underwear with a replaceable charcoal filter that removes bad-smelling gases before they escape. Weimer received a patent in 1998.

Buck Weimer and a pair of Under-Ease at the Ig Nobel Prize Ceremony. Photo: Jon Chase/Harvard News Office.

"The undies are made from a soft, airtight, nylon-type fabric. Elastic is sewn around the waist and both legs. The removable filter – which looks similar to the shoulder pads placed in women's clothing – is made of charcoal sandwiched between two layers of Australian sheep's wool."

Weimer began the research and development process by trying to modify a standard gas mask. When that proved unsuitable to the task, he began tinkering. The final design is a slick, low-tech device.

Biologists love hearing Weimer describe the action of the filter mechanism:

"The multi-layered filter pad traps the 1–2% of human gas creating the foul smell (mostly hydrogen sulfide) but allows the remaining non-smelling gas (mostly methane) to pass through. It also allows the natural build-up of body heat to pass through."

Engineers love hearing how it is constructed:

"A triangular 'exit hole' for the flatus to be expelled is cut from the back of the airtight underwear, near the bottom. This 'exit hole' is covered with a 'pocket' made of ordinary porous fabric sewn over the 'exit hole.' This unique design forces all expelled gas (flatus) out through the 'pocket.'"

Business gurus love hearing the company's motto: "Wear them for the ones you love."

Under-Ease comes in boxers for men, and panties for women. Replacement filters are available at low cost.

For easing the biological and social interactions of his fellow human beings, Buck Weimer won the 2001 Ig Nobel Prize in the field of Biology.

The winner and his wife traveled from Colorado, at their own expense, to attend the Ig Nobel Prize Ceremony. Buck Weimer presented pairs of Under-Ease to the Nobel Laureates, and instructed them on how to use them. In accepting the Prize, he said:

"My acceptance speech comes in the shape of a song. I think you all may remember it. It's called 'Imagine.'"

Imagine no odorous gas.
It's easy if you try:
No more burning nostrils,
Our tush securely covered.
Imagine all the people
Wearing Under-Ease.

You might say I'm a dreamer.
But I'm not the only one.
My wife is here also.
I hope someday you'll join us

With Under-Ease
the world will be one.

Imagine the odorless sound –
I wonder if you can.
No need for divorce or separation.
Free from shame and guilt.
Imagine all the people
sharing Under-Ease.

I'm almost finished ...

You may say I'm a dreamer,
but I'm not the only one.
My wife is still here.
I hope someday you'll join us,
with Under-Ease
the world will live as one.

THE UGLY: A Man Who Pricked His Finger and Smelled Putrid for 5 Years

"We ask assistance from colleagues who may have encountered a similar case or for suggestions to relieve this patient's odour."

–from the report by Mills, Llewelyn, Kelly, and Holt

THE OFFICIAL CITATION

THE IG NOBEL MEDICINE PRIZE WAS AWARDED TO

Patient Y and to his doctors, Caroline Mills, Meirion Llewelyn, David Kelly, and Peter Holt, of Royal Gwent Hospital, in Newport, Wales, for the cautionary medical report, "A Man Who Pricked His Finger and Smelled Putrid for 5 Years."

Their study was published in *The Lancet*, vol. 348, November 9, 1996, p. 1282.

Four doctors were hit nose-on with one of the most baffling medical mysteries on record. It all started with a chicken.

In September 1991, a 29-year-old man who dressed chickens for a living pricked his finger with a chicken bone. This fateful prick cause his finger to soon become reddish and smelly. The man got himself to the Royal Gwent Hospital, in Newport, Wales, where Drs Caroline Mills, Meirion Llewelyn, David Kelly, and Peter Holt took him under their care.

The doctors treated it with the antibiotic flucloxacillin. The hand still smelled.

Then they tried a different antibiotic, ciprofloxacin. The hand still smelled.

Next came erythromycin. Still the hand smelled.

Next up: metronidazole. The smell persisted.

The doctors delved into the hand surgically, but found nothing there of interest. They did a skin biopsy and cultured the microorganisms from it, hoping to discover some noxious bug. Here, too, they found nothing of interest.

Meanwhile, the man continued to stink.

The doctors took stool cultures. These stank, too, but only in the ordinary way.

The doctors tried everything they could think of: isotretinoin, psoralen; ultraviolet light treatment; colpermin; probanthene; chlorophyll; and even antibiotic withdrawal to allow restoration of normal flora. All to no avail. As they put it:

"Although the clinical appearance improved, the most disabling consequence of the infection was a putrid smell emanating from the affected arm, which could be detected across a large room, and when confined to a smaller examination room became almost intolerable."

At the five-year mark, the man still stank. At that point, the doctors wrote up a description of this curious case, and published it in hopes that some physician somewhere had encountered a similar problem and could suggest a way to relieve the patient's distress.

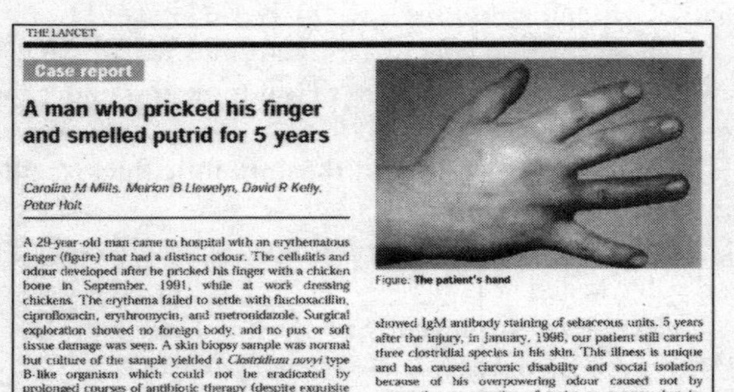
THE LANCET

Case report

A man who pricked his finger and smelled putrid for 5 years

Caroline M Mills, Meirion B Llewelyn, David R Kelly, Peter Holt

A 29-year-old man came to hospital with an erythematous finger (figure) that had a distinct odour. The cellulitis and odour developed after he pricked his finger with a chicken bone in September, 1991, while at work dressing chickens. The erythema failed to settle with flucloxacillin, ciprofloxacin, erythromycin, and metronidazole. Surgical exploration showed no foreign body, and no pus or soft tissue damage was seen. A skin biopsy sample was normal but culture of the sample yielded a *Clostridium novyi* type B-like organism which could not be eradicated by prolonged courses of antibiotic therapy (despite exquisite

Figure: **The patient's hand**

showed IgM antibody staining of sebaceous units. 5 years after the injury, in January, 1996, our patient still carried three clostridial species in his skin. This illness is unique and has caused chronic disability and social isolation because of his overpowering odour caused not by

The Prize-winning report.

For treating, and of necessity smelling, the unfortunate man who pricked his finger and smelled putrid for five years, Caroline Mills, Meirion Llewelyn, David Kelly, and Peter Holt – together with their unnamed, unfortunate patient – won the 1998 Ig Nobel Prize in the field of Medicine.

The winners could not, or would not, themselves travel to the Ig Nobel Prize Ceremony, but Dr Mills sent her cousin Matthew Edwards to accept on their behalf. This is the speech he read for them:

"Thank you for this award, in recognition of an extraordinary medical report. We published this case to seek help. Despite enormous amounts of correspondence, nobody had ever seen anything like this before, and no suggestions were effective. Our story, however, does have a happy ending. Our patient no longer smells putrid. Thank you very much."

Food – Preparation & Disposal

Many people simply consume food. Others devote much thought and effort to its preparation and disposal. Here are four examples of thoughts and efforts that led to Ig Nobel Prizes.

- Extremely Instant Barbecue
- Bright Blue, and Wiggly
- Salmonella Excretion in Joy-Riding Pigs
- Sogginess at Breakfast

Extremely Instant Barbecue

"'The fire department is real ticked, so I'm not going to do it any more,' he said. 'I told them that somebody could do the same thing with gasoline and that would make one heck of a fire and be a lot more dangerous, but I'm not going to argue with them.'

"Goble said West Lafayette fire officials threatened to cite him for violation of the city's open burning ordinance. 'I told them that if they did it, they'd have to cite everybody with a charcoal grill,' he said. But Goble said one fire prevention official was nice to him. 'He asked me if he could have a copy of a video of me lighting a fire. He said he wanted to use it for a training tape.'"

–from an article in the *Indianapolis* [Indiana] *Star*, October 6, 1996

THE OFFICIAL CITATION

THE IG NOBEL CHEMISTRY PRIZE WAS AWARDED TO

George Goble of Purdue University, for his blistering world record time for igniting a barbecue grill – three seconds, using charcoal and liquid oxygen.

George Goble's web site (http://ghg.ecn.purdue.edu/~ghg) features video of a quick grill ignition.

A computer engineer at Purdue University decided to optimize the process of igniting a barbecue grill. He succeeded.

George Goble likes barbecues. That is why he destroys them. The destruction is a mere side effect. The point of it all is speed.

The history of the project, like the ignition, is enlightening and brief. Goble once explained it to a local newspaper:

"Goble said he got the idea after cooking out with engineer pals for several years. 'It always took a half-hour to 40 minutes to get the thing going, so we started using hairdryers, a vacuum on low, and propane torches to get it going,'

George Goble ignites his lunch. Photo: Joe Cychosz.

Goble said. 'Then we took an oxygen tank like the kind scuba divers use and blew it through a ten-foot-long pipe. We were grilling in 30 seconds. Every year we got it started faster and faster until we got it down to a few seconds with so much pressure that it blew the briquettes out of the grill.'"

The optimized procedure is simplicity itself. Goble asks someone to throw a lighted cigarette on the grill. Then he pours three gallons of liquid oxygen onto the cigarette.

Safety is the number one, or, at worst, number two, concern. Goble pours the liquid oxygen from an eight-foot-long wooden pole. Spectators are kept at an almost reasonable distance. "Don't stare at the flame unless you squint," he advises them "It's like the sun."

While a cheap grill is entirely consumed by the fire, a sturdy make – a good Weber grill, for example, can survive two to three liquid oxygen barbecues before all trace of it vanishes.

Having achieved a consistent sub-four-second ignition time, and having attracted increasingly annoyed complaints from local fire officials, Goble has publicly stated that, in future, if he prepares ultra-fast food, it will not be in this particular manner.

For establishing a new standard for barbecue grill ignition, George Goble won the 1996 Ig Nobel Prize in the field of Chemistry.

The winner could not travel to the Ig Nobel Prize Ceremony, but instead sent his colleague Joe Cychosz to accept on his behalf. Mr Cychosz (who pronounces his name, with lifelong matter-of-factness, like the plural of "the word psycho") said:

"It's hard to believe all that work started back with a hairdryer and a bunch of people who wanted to eat real quick. On behalf of George, I'd like to thank the Board of Governors."

Bright Blue, and Wiggly

"To understand the genius of Jell-O gelatin, you have to understand how gelatin desserts were originally made. First you had to get two calves' feet – scald them, take off the hair, slit them in two, and extract the fat from between the claws. Then you had to boil them, remove the scum, and boil them again for as long as six or seven hours – before straining them, letting them cool, skimming the fat, boiling once more, adding the shells and whites of five eggs (to pick up impurities), skimming again, and straining twice through a jelly bag that you would have to make yourself. Then you would add flavoring, sugar, and spices; pour into a jelly mold; pack with ice; and go to bed while it sets – it now being midnight."

–from the book *Jell-O: A Biography*

THE OFFICIAL CITATION

THE IG NOBEL CHEMISTRY PRIZE WAS AWARDED TO

Ivette Bassa, constructor of colorful colloids, for her role in the crowning achievement of 20th-century chemistry, the synthesis of bright blue Jell-O.

The book *Jell-O: A Biography*, by Carolyn Wyman, Harcourt, 2001, is a comprehensive study of Jell-O.

In 1992, in a shocking break with tradition, Kraft General Foods, the manufacturer of Jell-O brand gelatin, introduced a version of the dessert that was intentionally gross and disgusting.

Jello-O brand gelatin has long called itself "America's Most Famous Dessert." It was generally considered a comfort food and a fun food. For nearly 100 years, the manufacturer tried to offer it in a variety of tempting colors. The emphasis was on "tempting," and most of the colors were – at least in theory – the same colors to be seen in some of nature's most temptingly delicious foods: raspberry, strawberry, lemon, cherry, orange, banana, and the like.

Then, in 1992, the company tried a simple experiment. They created a new variety of Jell-O with a color specifically calibrated to disgust and nauseate adults, the theory being that children would find the adults' reaction disgustingly and nauseatingly and irresistibly enchanting.

The experiment was an instant, enormous success. Called "Berry Blue," it was a glowing, aliens-invading-in-flying-saucers shade of blue, and it became the third highest-selling variety of Jell-O.

The initially unsung hero of the Berry Blue Jell-O story was a food technologist named Yvette Bassa, who figured out how to chemically construct Jell-O of that color and still have it taste good.

For her exceedingly colorful contribution to science and nutrition, Yvette Bassa won the 1992 Ig Nobel Prize in the field of Chemistry.

The Prize-winning variety of Jell-O.

Yvette Bassa and an entire team from Kraft General Foods traveled to the Ig Nobel Prize Ceremony, at their own expense, on their company's corporate jet. In accepting the Prize, Yvette Bassa said:

"I feel humbled at being singled out for this honor. My achievement is simply the capstone on an immense body of scientific research performed over the past hundred years. My colleagues and I wish to direct your attention to three great chemists who laid the foundations for our work:

"Emil Fischer, who won the Nobel Chemistry Prize in 1902 for synthesizing sugars and purine derivatives; Richard A. Zsigmondy who won the 1925 Nobel Prize for his pioneering work in colloidal chemistry; and Linus Pauling, who won the 1954 Nobel Prize for his discoveries about the nature of chemical bonds. Most of all, I would like to acknowledge our debt to the great 19th-century scientist whose research opened the door to 20th-century chemistry – Pearl B. Wait, a manufacturer of cough medicine in Leroy, New York, who in 1897 became the first person to synthesize Jell-O gelatin dessert."

After the ceremony, Bassa and her team passed out freshly-made bright blue Jell-O to the audience. All the children wanted it; most of the adults emphatically did not.

Salmonella Excretion in Joy-Riding Pigs

"We have called it the experiment of the joy-riding pigs because it is based upon a confidence trick played upon a pig population."

–from Williams and Newell's report

THE OFFICIAL CITATION

THE IG NOBEL BIOLOGY PRIZE WAS AWARDED TO

Paul Williams Jr of the Oregon State Health Division and Kenneth W. Newell of the Liverpool School of Tropical Medicine, England, bold biological detectives, for their pioneering study, "Salmonella Excretion in Joy-Riding Pigs."

Their study was published in *American Journal of Public Health and the Nation's Health*, vol. 60, no. 5, May, 1970, pp. 926–9.

Two public health specialists resorted to trickery in their efforts to solve a medical mystery stemming from a perpetually pressing problem with pigs.

The problem was: How do you keep people from getting salmonella? The answer, decided Paul Williams and Kenneth Newell, might be obtained by sending pigs on a joy ride.

Salmonella is a nasty and most unpleasant bacterium. It takes just a tiny amount of it to infect a human. The infection brings, just a few hours later, nausea, vomiting, abdominal cramps, diarrhea, fever, and headache, with additional delights possibly to follow. The repulsive little salmonella bugs are commonly found in poultry and swine, who commonly get it from contaminated water, soil, or feed.

The puzzle was: Pigs were regularly tested for salmonella while they were at the farm. They would get a clean bill of health. But, after their short journey to the slaughterhouse, anywhere from 30 to 80% of them would test positive for salmonella.

The salmonella test for a pig consisted of taking a rectal swap, and then analyzing it in the lab.

Some scientists thought the pigs were getting salmonella from something – maybe from other pigs, maybe from the environment – in the truck or in the holding areas. Other scientists thought maybe the pigs already had salmonella when they were at the farm, but it was deep within them, and the stress of their fateful final journey would cause the infection to pass through the intestines and thence into the rear exit area, at which location it would be show up on the swab test.

The Williams-Newell experiment, which Williams and Newell called "the experiment of the joy-riding pigs," was designed to test the latter possibility. They described it in simple terms:

"Using rectal swabs, we would examine this group of pigs on the farm and find their salmonellae excretion rate by this method. We would then take a truck that had been cleaned and made free from salmonellae, and load these pigs upon it as if they were going to be slaughtered. However, we would give them a joy ride through the countryside and end the trip, not at the slaughterhouse, but back at the farm. We would examine them again at this time."

Williams and Newell report that prior to the experiment, "the pigs were fed stale bread, rolls, cakes, and other bakery items (including their wrappers) which had been run through a hammer-mill." (In his Ig Nobel Prize acceptance speech, Newell added that "bran was added to keep the hammer-milled bread and roll wrappers moving through the porkers.")

The joy ride itself was uneventful:

"The day of the experiment was a pleasant, clear one

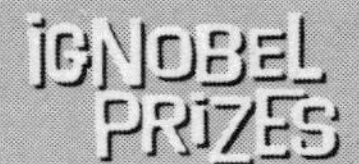

JOY-RIDING PIGS ... AND YOU!

In their Ig Nobel Prize-winning paper, Williams and Newell point out that the joy-riding pigs experiment has implications for human beings:

"If other animals and man behave in a similar way to these pigs under study, then the evidence we use to judge the pathogenicity of enterobacteriaceac and the relationship of enteric infection to gastrointestinal illness may also be invalid."

In other words, if your doctor does a stool test on you and the test says you are A-OK, the test could be missing something.

Your doctor may not be aware of this discovery. We recommend that next time you go in for a check-up, you would do well to bring a copy of Williams and Newell's report with you. In particular, we recommend that you draw your doctor's attention to the final section of the report, where it says:

"The hypothesis that man and the agents of man may behave in this way is supported by the work of Gangarosa and his co-workers. These authors showed that the judgment of human excretion of Vibrio cholerae, biotype El Tor, could be altered by a 'preparatory purge.'"

with a temperature of 70–78 degrees F ... The pigs were loaded on the truck and were given a 60-mile joy ride. The truck was then parked in the shade for half an hour, and later was taken for another 90 miles before the pigs were returned to the farm. The total joy ride lasted for 3¾ hours. On their return, the pigs were once again rectally swabbed, and six pigs (30%) were shown to be excreting salmonellae."

And so they had done it. Paul Williams, and Kenneth Newell had demonstrated conclusively that joy-riding pigs excrete salmonella. For this achievement, they won the 1993 Ig Nobel Prize in the field of Biology.

Kenneth Newell passed away in 1990. Paul Williams could not travel to the Ig Nobel Prize Ceremony, but sent an audio-taped acceptance speech, in which he said:

"Dr Newell coined the catchy – but appropriate – title for the article reporting this work. The experiments and the article were not whimsical, albeit they were a little unusual. Whatever, I humbly accept this great, and somewhat dubious, honor for Dr Newell and me, for what it is worth. It is good that we did the experiments in 1967, because they could not be duplicated today. Animal rights extremists would likely come out of the woodwork, firmly believing that it was inhumane to play confidence tricks on pigs – especially ones being fattened on bakery goods originally intended for human consumption."

The following year, the chairman of the Ig Nobel Board of Governors formally presented the Prize to Paul Williams at a special public ceremony in Portland, Oregon, in Powell's book store, before an unexpectedly large crowd of joy-riding-pig-experiment enthusiasts.

Sogginess at Breakfast

"Wheat breakfast flakes were compacted in a cylindrical geometry using two different techniques and the volume measured as a function of applied pressure from 100 Pa to 85 Mpa. The effect of water content, in the range 4 to 18% (wet weight basis), on the compaction behaviour of the flakes was examined for pressures from 1 to 85 Mpa ... The Heckel deformation stress decreased as the water content increased up to 12% and became inaccurate at 18% water. The Peleg compressibility and Kawakita yield stress only showed a marked decrease from 12 to 18% water."

–from Georget, Parker, and Smith's report

THE OFFICIAL CITATION

THE IG NOBEL PHYSICS PRIZE WAS AWARDED TO

D.M.R. Georget, R. Parker, and A.C. Smith, of the Institute of Food Research, Norwich, England, for their rigorous analysis of soggy breakfast cereal, published in the report entitled "A Study of the Effects of Water Content on the Compaction Behaviour of Breakfast Cereal Flakes."

Their study was published in *Powder Technology*, November, 1994, vol. 81, no. 2, pp. 189–96.

Many people, of a morning, wonder why their breakfast cereal becomes soggy. Thanks to a painstaking investigation conducted in 1994, the answer is now on the public record.

D.M.R. Georget, Roger Parker, and Andrew Smith looked at the basic physics of a breakfast cereal flake.

They examined the flake with fresh eyes, augmented by the obvious battery of equipment: a Mettler LP16 moisture balance; an Instron 1122 Universal testing machine; a piston-driven capillary rheometer; and all the rest.

Prior to this time, anyone who wanted to understand the basic physics of a breakfast cereal flake had to camp out in engineering libraries and dig through the many reports that touched on this or that aspect of breakfast-cereal-flake behavior.

Georget, Parker, and Smith did things methodically. Before touching their cereal, the three chums devoured the published works of all the great breakfast-cereal-flake researcher teams: Peleg, Kawakita and Heckel; Roberts and Rowe; Train and York; Illka and Paronen; and the never-to-be-forgotten Marousis and Saravacos.

Once they knew what was known and what was not, Georget, Parker, and Smith got right down to business. They obtained breakfast cereal flakes. They performed experiments. They did calculations. They plotted plots and graphed graphs. And ultimately, they solved the puzzle.

Basically, they soaked some flakes in water, then dropped them into a cylinder, and then stuck a thingy down into the cylinder to compress the flakes. They measured how much the flakes compressed as they got soggier and soggier. To get a thorough picture of the sogging process, they did this again and again, each time using slightly soggier flakes.

They discovered that, up to a point, as a flake takes on liquid, it retains much of its girlish firmness. But after that point, it goes suddenly limp. To put this in simple language: the Heckel deformation stress becomes increasingly sensitive to the particle density as the water content increased. That may seem obvious now, but at the time it was only fairly obvious.

The journey from crisp to soggy is considerably more colorful than people expect, especially in a numerical sense. For example: the biggest changes in sogginess come as the water content of the flake increases from 12% to 18%. But the fun is in crunching all the numbers, so the reader is urged to get a copy of Georget, Parker, and Smith's full report, and also perhaps get a bowl of cereal, and sit down for a multidimensional, crackling good feast of the senses.

A word of caution, though – Georget, Parker, and Smith obtained all their results using water. In theory, their results will hold up when, some day, someone repeats the experiments using milk. For the time being, the story seems to hold water.

Georget, Parker, and Smith's "Study of the Effects of Water Content on the Compaction Behaviour of Breakfast Cereal Flakes" is a high point in the intellectual history of cereal-flake soggification. For advancing our basic understanding of what happens in the bowl, D.M.R. Georget, Roger Parker, and Andrew Smith won the 1995 Ig Nobel Prize in the field of Physics.

The winners could not travel to the Ig Nobel Prize Ceremony. Instead they sent a videotaped acceptance speech of themselves in their lab, with a bowl of cereal. In the video, Andrew Smith spoke for the three of them:

"In our study of compaction of breakfast cereal flakes, we did not leave them turned tongue-in-cheek, or use any other sensory technique. Rather, we set out to relate macroscale mechanical properties to changes in the scale of constituent food particle molecules. This provides valuable insights into texture. So what does this mean for the manufacturer, and to you, the consumer? Well, it's all about the quest for the ultimate breakfast-cereal-eating experience. I hope that the awarding of this prize will stimulate further research in this area."

NOTE: News of this particular Prize indirectly touched off the curious affair of the Chief Science Adviser's Angry Complaint. See the introductory section of this book for details.

Food – Palatability

There is an old saying, "I don't live to eat – I eat to live." Ig Nobel Prizes have been won for doing experiments to prove or disprove every part of that statement. This chapter describes three of those endeavors:

- The Comparative Palatability of Tadpoles
- The Effects of Ale, Garlic, and Soured Cream on the Appetite of Leeches
- No Need for Food

The Comparative Palatability of Tadpoles

"Hopefully, this will some day be verified with a larger sample of tadpoles."

–from the concluding paragraph of Richard Wassersug's research report

THE OFFICIAL CITATION

THE IG NOBEL BIOLOGY PRIZE WAS AWARDED TO

Richard Wassersug of Dalhousie University, for his first-hand report, "On the Comparative Palatability of Some Dry-Season Tadpoles from Costa Rica."

The study was published in The American Midland Naturalist, vol. 86, no. 1, July, 1971, pp. 101–9.

Most scientific reports begin with an abstract, a little summary that by tradition is dry and unappetizing. The abstract of Richard Wassersug's 1971 report is dry, certainly, but it does whet the reader's reading appetite. It says:

"ABSTRACT: Tadpoles of eight species of frog were tasted in a standard procedure by 11 volunteers. The tadpoles were rated in their palatability from 'tastes good' to 'highly unpalatable.' It is suggested that palatability in tadpoles may correlate inversely with vulnerability."

Why would a scientist eat a tadpole – or rather, why would a scientist talk 11 other scientists into eating tadpoles? To answer a scientific mystery, of course; and also because he could.

Tadpoles come in an almost unbelievable variety of patterns and colors. Most blend in with the sand, rocks, or vegetation of the stream or pool where they live. But some tadpoles are covered with gaudy patterns or bright colors, or both. Why don't predators eat them? They are so easy to spot that you might expect every one would be gobbled up long before it could metamorphose into a frog – and so that kind of frog, and that kind of tadpole, would almost instantly become extinct.

The leading theory was that the eye-catching tadpoles must taste terrible to predators, so yucky that predators spurn them. But this was, as they say, "just a theory" – until Richard Wassersug came up with a way to test it out. He did so in Costa Rica, where tadpole species are numerous, and beer to wash them down with is cheap.

Tadpoles have been described as "nothing more than miniature, underwater cows that graze on algae." Generally speaking, if something wants to eat a tadpole, it can. And lots of somethings *do* want to eat tadpoles – beetles, water bugs, dragonflies, many kinds of fish, and pretty much anything big enough to fit one inside its maw. But one can't ask these hungry creatures to sit at a table and conduct a taste test in gentlemanly fashion. Thus, Richard Wassersug decided, he would use substitutes for the natural predators of the tadpole. He would use cheap substitutes. He would use graduate students.

Wassersug laid out strict procedures.

Every effort was made to collect only tadpoles of equal size. The tadpoles were kept alive in clear, fresh water for several hours before the test began.

Here is Wassersug's description of the taste test:

"The experiment was run at least 2½ hours since the last meal for the volunteer tasters. Each taster was tested separately and asked not to discuss the test until the experiment was over. The tadpoles, which were assigned a number, were presented individually to each subject one at a time and by number rather than name. The taster did not know which tadpole had which number ...

"The tasters were asked to rate the palatability of each

tadpole's skin, tail, and body on a 1 to 5 scale: 1, tastes good; 2, no taste; 3, only slightly disagreeable; 4, moderately disagreeable; and 5, very strongly disagreeable. They were also asked to make comments about the taste as they went along and to note the most and least palatable tadpole at the end of the experiment. The standardized tasting procedure included several steps. A tadpole was rinsed in fresh water. The taster placed the tadpole into his or her mouth, and held it for 10–20 seconds without biting into it. Then the taster bit into the tail, breaking the skin and chewed lightly for 10–20 seconds. For the last 10–20 seconds, the taster bit firmly and fully into the body of the tadpole. The participants were directed not to swallow the tadpoles, but to spit them out and to rinse their mouths out at least twice with fresh water before proceeding to the next tadpole."

Wassersug eliminated two of the 11 tasters because they were heavy cigarette smokers who had difficulty tasting anything. The remaining nine performed with a professional level of competence.

The results of the experiment were sparklingly clear. One species of tadpole, Bufalo marinus, was rated the most distasteful by six of the nine tasters. It was described as being "bitter," and nobody professed a liking for it. Several species were found to taste almost good. Generally, the tasters found the bodies to be less palatable than the skin, but more palatable than the tails.

On the whole, the results supported Wassersug's thesis that the more conspicuous the tadpole, the less palatable it is likely to be. For achieving this small but stimulating advance in scientific knowledge, Richard Wassersug won the 2000 Ig Nobel Prize in the field of Biology.

At the time he conducted this experiment, Richard Wassersug was based at the University of California at

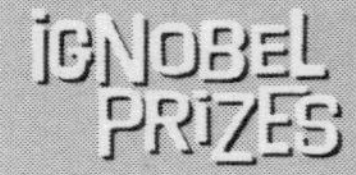

OTHER WASSERSUGIAN RESEARCH

Richard Wassersug has always had a taste for the unusual. Throughout his career as a biologist, he has conducted research on a wide variety of questions, mostly of great scientific import. Some of his experiments have been more colorful than others. Among them:

- 1993. Wassersug and a colleague published a report titled "The Behavioral Reactions of a Snake and a Turtle to Abrupt Decreases in Gravity."
- 1993. In a report titled "Motion Sickness in Amphibians," Wassersug and two colleagues "explored the question of whether amphibians get motion sickness by exposing anurans (frogs) and urodeles (salamanders) to the provocative stimulus of parabolic aircraft flight. Animals were fed before flight, and the presence of vomitus in their containers after flight was used to indicate motion-induced emesis [vomiting]."
- 1996. Japanese tree frogs (Hyla japonica) were exposed to 35 cycles of altered gravity on the FreeFall "G.0" ride at the Space World amusement park in Kitakyushu, Japan. The results "bode well for the potential of anurans to breed in microgravity and to be used for biological research in space."

Berkeley. 29 years later he was a tenured professor of biology at Dalhousie University in Halifax, Nova Scotia. He traveled from there to the Ig Nobel Prize Ceremony, at his own expense. In accepting the Prize, he said:

"Finally, an Ig Nobel Prize for a paper that is not just another piece of tasteless research. I want to point out, however, that the reference to "dry season" tadpoles in the title refers to when the tadpoles were found. It is not a reference to how the tadpoles were prepared – they were neither dried nor seasoned.

"I couldn't have won this award without the immense help of all my fellow graduate students in the jungle biology course that I took in Costa Rica 30 years ago. My compatriots, who willingly chewed up tadpoles for the reward of a few beers, are heroes to me. I thank them all. And we all thank the Organization for Tropical Studies for the opportunity to do this research and to drink Costa Rican beer.

"I also thank Hugh Cott, who, before me, pioneered the gentlemanly science of tasting wildlife from strange places in an effort to help British soldiers stranded in exotic countries during World War II. It was a great thing Cott did when he organized the Egg Tasting Panel of British elders and proved that birds with palatable eggs were smart enough to put those eggs on the sides of horrendously steep cliffs where no British soldier could safely get to them."

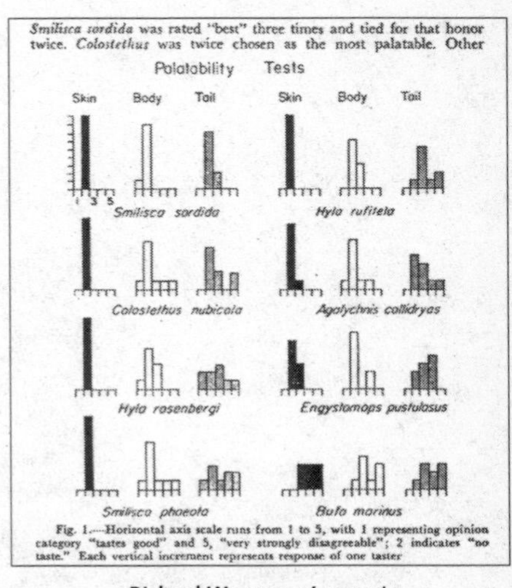
Smilisca sordida was rated "best" three times and tied for that honor twice. *Colostethus* was twice chosen as the most palatable. Other

Fig. 1.—Horizontal axis scale runs from 1 to 5, with 1 representing opinion category "tastes good" and 5, "very strongly disagreeable"; 2 indicates "no taste." Each vertical increment represents response of one taster

Richard Wassersug's report.

The Ig Nobel Board of Governors attempted to organize a tadpole taste test to celebrate Professor Wassersug's appearance two days later at the Ig Informal Lectures, held at MIT. The effort was stymied by a combination of federal laws and university regulations, and a paucity of volunteers.

The Effects of Ale, Garlic, and Soured Cream on the Appetite of Leeches

"The medicinal leech has regained some of its lost popularity by its present use in microsurgery. Sometimes, however, the leeches refuse to cooperate properly. To overcome this problem doctors in the 19th century used to immerse leeches in strong beer before applying them to the patient. In the 1920s a deaconess experienced that a little soured cream on the skin would encourage the leeches' feeding behaviour, and we recently found that they seem to be attracted by garlic. We designed a study to evaluate the effect of these remedies."

–from Baerheim and Sandvik's report in the *British Medical Journal*

THE OFFICIAL CITATION

THE IG NOBEL BIOLOGY PRIZE WAS AWARDED TO:

Anders Baerheim and Hogne Sandvik of the University of Bergen, Norway, for their tasty and tasteful report, "Effect of Ale, Garlic, and Soured Cream on the Appetite of Leeches."

Their study was published in the *British Medical Journal*, vol. 309, Dec 24–31, 1994, p. 1689.

How, exactly, does one stimulate the appetite of a leech?

Until the mid 20th century, leeches were a common tool in medicine. Recently, after several generations of disuse, they have made a rather glamorous return to the medical scene. Surgeons who perform certain kinds of microsurgery clamor for leeches. In reattaching a severed finger, for example, it is crucial to keep the blood from clotting – and applying a leech is far and away the best known way to do that.

A hungry leech is a welcome addition to a surgical team. Yet leeches, like most humans, are not reliable trenchermen. And in a surgical emergency, a sated, passive leech is of no use.

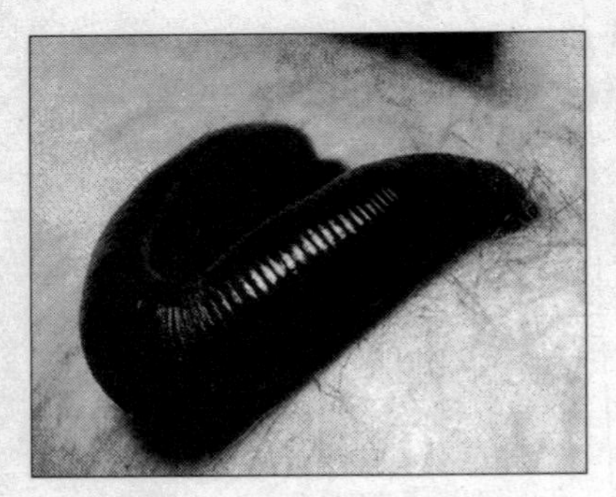

A leech on an arm during the test procedure. This photo is from the original *British Medical Journal* report.

Then, how does one stimulate the appetite of a leech? Chances are your doctor would rely on the wisdom passed down from medical authorities of the 19th and early 20th centuries: that beer or soured cream are guaranteed to give even the most bloated leech a powerful case of the munchies.

Until quite recently, medicine was very much an art, and almost not at all a science. In 1994, Anders Baerheim and Hogne Sandvik realized that no one had *scientifically* tested the conventional methods of leech appetite stimulation. This, therefore, was what they set out to do. And, in addition to testing the traditional leech treats – beer and soured cream – they added a new delight: garlic.

Baerheim and Sandvik used a simple laboratory procedure:

"Six leeches were dipped briefly in one of two different types of beer (Guinness stout or Hansa bock) ... before being placed on the forearm of one of us (HS). We measured the time from when the leech touched the skin until HS felt it bite. Each leech was exposed three times to each liquid in random order."

The experiment yielded clear results:

BEER: "After exposure to beer, some of the leeches changed behaviour, swaying their forebodies, losing grip, or falling on their backs."

GARLIC: "Two leeches placed on the forearm smeared with garlic started to wriggle and crawl without assuming the sucking position ... Their condition deteriorated. When placed on a bare arm they tried to initiate feeding but did not manage to coordinate the process. Both died 2½ hours after the exposure to garlic. For ethical reasons the garlic arm was abandoned at this point."

SOURED CREAM: Leeches exposed to soured cream became ravenous. When they were then placed into a glass beaker, they "sucked frantically on the wall of their container after they had been on the arm." While they were still on the arm, however, they bit no sooner than leeches that had been deprived of sour cream or any other artificial stimulant.

So, how does one stimulate the appetite of a leech? Science does not really know. But best not with beer or garlic, and likely not with soured cream.

Anders Baerheim and Hogne Sandvik scientifically – and boldly – tested the medical profession's best accepted "wisdom." They demonstrated that the conventional wisdom was wrong. For this they were awarded the 1996 Ig Nobel Prize in the field of Biology.

The winners could not attend the Ig Nobel Prize Ceremony, but instead sent a videotaped acceptance speech:

"We accept this prize with profound gratitude. We accept it as a tribute to our partners, the leeches, who showed remarkable enthusiasm during our experiments. Lab animals seldom receive credit for their scientific achievements. The leeches' reaction to this honor was predictable – they celebrated. Since the leeches are in no fit state to accept this award today, we have to rely on a stand-in. As he is not a leech, we are confident he will behave with proper dignity."

The mysterious "stand-in" turned out to be a tall,

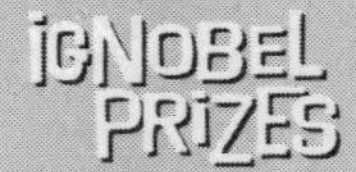

KEEP A FINGER ON THE LATEST RESEARCH

Baerhem and Sandvik's discovery could prove important – very important – to you personally. We recommend that you photocopy this report and keep a copy with you at all times. Here's why.

Should you be misfortunate enough to lose a finger, but fortunate enough to find that finger and bring it to a hospital, you and your digit will be taken to an operating room. The surgeon and nurses will arrive. Their instruments will be arrayed. The operation will begin. To keep the blood in your severed finger and in your stumped knuckle from coagulating, they will fetch a leech. All's well, so far.

But what if the leech happens not to be hungry? What will the doctor do then?

Should your surgeon not be up on the latest medicinal leech research, she or he might attempt to stimulate the leech's appetite with beer or with garlic, thus rendering the animal inebriate or dead. The reunion of you and your finger would not occur.

But if you have diplomatically educated your surgeon by showing them this printed documentation, the medical staff will prep the leech properly – shunning beer and garlic – and your finger reattachment surgery will proceed rapidly to a happy result.

somber-faced gentleman named Terje Korsnes, Norway's Honorary Consul to Massachusetts. Korsnes stepped to the lectern, delivered a brief speech to the 1,200 people in the packed theater, and then he delivered something else. What he said was:

"Thank you. My countrymen Baerheim and Sandvik couldn't physically make it to the ceremony. Clearly, this kind of breakthrough research should be recognized. I'm sure that this is the kind of recognition that the scientific community in Bergen has been waiting for. I do fear that this topic is not taken seriously here in Cambridge. Perhaps in this audience with bright minds there is someone who would like to leech onto this topic for further research. To

IG NOBEL PRIZES

NOT ALL SCIENTISTS ARE SHY

Anders Baerheim and Hogne Sandvik are unique in the ranks of Ig Nobel Prize winners.

After publishing "Effect of Ale, Garlic, and Soured Cream on the Appetite of Leeches" in the prestigious *British Medical Journal*, they sent a copy of it to the Ig Nobel Board of Governors, accompanied by a note explaining why they believed themselves deserving of an Ig Nobel Prize.

In 1996, they were selected to win one of that year's crop of Igs. They triumphed against a field of several thousand nominees, several hundred of whom had nominated themselves.

Although every year many people nominate themselves for an Ig Nobel Prize, of the more than 100 Prizes awarded so far, the team of Baerheim and Sandvik is the only self-nominated winner.

facilitate that, I have brought these leeches to distribute among you so you can start your project."

At that point he reached into his suit pocket, removed a bag full of leeches, and began throwing them into the audience. The leeches were made of plastic, but the Ig Nobel organizers had no way of knowing that – and the audience *certainly* had no way of knowing it.

And so, on that day, the world learned something about leeches and about medicine, and it also learned something about Norwegian diplomats.

No Need for Food

"Nor is it a process of denial. During the last years I have often been guided by the *Masters* to even stop drinking fluids. They have assured me that the body needs only 'Liquid Light.' But, I like socializing over a cup of tea with friends and – at the time of writing – I have not yet conquered my intermittent boredom with lack of flavour."

–Jasmuheen, in the foreword to her book *Living on Light*

THE OFFICIAL CITATION

THE IG NOBEL LITERATURE PRIZE WAS AWARDED TO

Jasmuheen (formerly known as Ellen Greve) of Australia, first lady of Breatharianism, for her book *Living on Light*, which explains that, although some people do eat food, they don't ever really need to.

Jasmuheen described her nutritional history and outlook in the book *Living on Light*, KOHA Publishing, 1998.

The Breatharians are a loose-knit, but jolly, group who say they enjoy not eating together, and also enjoy not eating alone. The basic fact of Breatharianism is that they say they don't eat at all: and if they do eat, it is for social reasons or for entertainment, not for nourishment.

Jasmuheen, whose original name is Ellen Greve, says that she last ate a proper meal in 1993. Her sustenance and vitality since that time has come from "tapping into an alternative source of nourishment." Her diligence and lack of diet have made her the best-known Breatharian.

What is Breatharianism? To those who attend her seminars or buy her books, Jasmuheen offers a clear, simple definition of the practice:

"Breatharianism is the ability to absorb all the nutrients, vitamins, and nourishment one requires, to maintain a healthy physical vehicle, from the universal life force, or chi energy. A being who practices this does not need to eat food."

Jasmuheen practices this, and does not need to eat food. As reported on the back cover of her best-selling book *Living on Light*, "Since 1993, Jasmuheen has been physically nourished by the Universal Life Force of Prana."

Despite her obvious glowing good health and high spirits, Jasmuheen has been afflicted by quibblers and doubters. A report in *The Times* of London, on April 6, 2000, for example, said that an Australian journalist who was checking onto a flight with Jasmuheen was surprised to hear the airline attendant ask the cult leader to confirm that she had ordered a vegetarian meal. Jasmuheen quickly denied it, then changed her mind. " 'Yes, I did, but I won't be eating it,' she said."

Jasmuheen is gracious in dealing with skeptics. On her web site she addressed this very incident:

"Fact: a *Daily Mirror* reporter overheard a flight-check-in lady confirming a 10-year standing note on my ticket for vegetarian food that I have never bothered to change. Fact: on long flights I sometimes eat a potato to send me to sleep as when I never eat I find it very hard to sleep. Eating a small amount can activate the digestive process, drops our energy levels and lets us sleep. No big deal."

The year before, the Australian news program *60 Minutes* arranged a test of Jasmuheen's claims. They sealed her into a hotel room in Brisbane for seven days, with a team of reporters watching closely to see that she ingested no food, and a doctor on hand to monitor her health. On the third day, the doctor expressed concern that she was showing signs of dehydration and physical distress, and so Jasmuheen and the entire group moved to a pleasant, and presumably less stressful, mountain retreat just outside the city. Two days later, *60 Minutes* halted the test and rushed Jasmuheen to a hospital.

In this case, Jasmuheen answered the critics, as she always does, with her trademark tact and honesty. Her organization issued a press release saying:

"*60 Minutes* decided to not continue the Challenge with Jasmuheen as she entered day 5 of the 7 days, on the advice of the attending Doctor, Dr Wenck. As Dr Wenck had not read Jasmuheen's in-depth research into this matter, the doctor became most concerned by Jasmuheen's weight loss and slight dehydration as would any doctor who was unfamiliar with the last 7 years of research into this matter. As Jasmuheen told *60 Minutes* before they began, it was imperative that they read the book detailing her research on this, so that they could be well informed as to the changes that would occur within Jasmuheen's system and not panic. Despite this, the producer, Richard Carleton and Dr Wenck admitted to Jasmuheen that they had not read this literature and hence, when they found themselves in what they felt was a potentially dangerous situation, they pulled the plug even though Jasmuheen was happy and able to safely continue."

Despite the doubters and the cynics, Jasmuheen has prospered. Her organization, the Cosmic Internet Academy, which is also known as the CIA, and which is dedicated to One People in Harmony to the Planet, distributes educational materials and arranges seminars and retreats. Reportedly, her books (*Pranic Nourishment, In Resonance, Inspirations* (Volumes I, II & III), *Streams of Consciousness* (Volumes I, II, & III), *Our Camelot – the Game of DA, Ambassadors of Light, Living on Light, Our Progeny – the X-re-Generation*, and *Dancing with My DOW*) and tapes (*Australia Overseas, Breath of Life, Inner Sanctuary, Emotional Realignment, Meditation for Empowerment, Ascension Acceleration, Self-Healing*

Meditation, and *Akashic Records Meditation*), her five-day Luscious Lifestyles Retreats, and her seven-day MAPS (Movement of an Awakened Positive Society) International Retreats all sell well, requiring her to keep a busy international schedule.

People constantly ask Jasmuheen whether she lost weight after giving up food. The answer is yes: "I programmed my body to reach a certain weight and stabilize there. I maintained a weight of 47–48 kilos from that time regardless of how much fluid I drank or how many intermittent mouthfuls of flavor I experimented with."

She goes on to explain that "Trying to gain weight after the process is more difficult than simply addressing the underlying belief patterns and not losing it in the first place!" One year, despite the tremendous difficulty, she did manage to gain eight kilos.

Although Breatharians can achieve nearly complete control over all bodily functions, Jasmuheen counsels that sometimes there are social reasons to do otherwise. For example, "Regarding menstruation, as I have completed my child-bearing, I thought it would be convenient to program for the body to cease bleeding. When this did not work I asked the Masters for guidance and was told that regular menstruation was a 'traditionally' acceptable sign of a healthy body and it was, and would be, important to continue to display signs of complete health."

On Jasmuheen's web site (www.jasmuheen.com) she says that her top priority is "to healthily feed, clothe, comfortably shelter and holistically educate the earth's people by 2012."

For her efforts to feed things to the world, Jasmuheen won the 2000 Ig Nobel Prize in the field of Literature.

The winner could not, or would not, attend the Ig Nobel

Prize Ceremony, but did carry on an engaging e-mail conversation with the Ig Nobel Board of Governors. The discussion was wide-ranging, not so much in topic as in location. Jasmuheen was traveling from country to country, conducting seminars and sampling the local light. She expressed pride and delight at winning the Ig Nobel Prize, and regret at having a previously scheduled, very lucrative event in Brazil that would preclude her coming to Harvard.

IG NOBEL PRIZES REMEMBRANCE: OF THINGS NOT EATEN

Jasmuheen is not the original Breatharian, nor does she claim to be. "Breatharianism," she writes in her book, "has been around since the dawn of time."

Here is a list of other notables who, early on, Just Said No to food.

This information was compiled by Juergen Buche, a naturopath at the Preventorium Institute in Montreal, who found it "in the press and in esoteric publications." Much of the wording here is Buche's; the sources are as he lists them.

- JUDAH MEHLER, Grand Rabbi, 1660–1751, "ate and drank sparingly one day a week, broke his 'fast' about twelve times a year on Jewish holidays, led a busy life as Rabbi of three communities and lived to be 91." (Source: *Ripley's Believe It or Not*.)
- MARIE FRUTNER, a Bavarian girl, "lived on water without food for 40 years and was under observation for a time in Munich in 1835."(Source: Hilton Hotema of *Health Research*.)
- YAND MEL, age 20, "who hasn't eaten for the last nine years: She shows no signs of starvation, leads a perfectly normal life except for having lost [the] desire for food." Her alimentary tract has become dormant and rudimentary; she takes no water. (Source: Dr T.Y. Ga, as mentioned in Jones, H. B. et al, *Am. J. Cancer*, 40: pp. 243–50, 1940.)
- GIRI BALA of Bahar, West Bengal, "now over 70 years old, who, as a child, had an insatiable appetite but has taken no food nor fluid since she was 12. She has never been sick, is an expert in pranayama (breathing exercises) and yoga, is always happy, looks like a child, does normal housework and has no bodily excretions. Her case was investigated by the late Sri Bijali Chand Mahtab, Maharajah of Burdwan, India." (Source: Paramhansa Yogananda, in his book *Autobiography of a Yogi*, 1946.)
- DANALAK SHUMI of Marcara, India, age 18, "for over one year took no food or water. She leads a normal healthy life. At age 14, her appetite diminished until she could not assimilate anything. The Indian Government sent her to be examined at the Bangalore General Hospital." (Source: the *Bombay Press*, August 1953.)
- BALAYOGINI SARASVATI of Amma, India, "lived on water only for a period of more than three years." (Source: the *Rosicrucian Digest*, June 1959.)
- CARIBALA DASSI, "sister of Babulamboxer, Pleader of Purillia, has been living for the past 40 years without taking any food or water and has done regular household duties with no apparent injury to her health, as per India's *Message*, 1932." (Source: none given.)
- TERESA AVILA, "a Bavarian peasant, born 1898, has taken no food nor water and does not sleep since 1926. She is not thin nor sickly, works in her garden and is described by 'Aberee 1960' as one of the happiest persons." (Source: none given.)
- THERESE NEUMANN, "a German nun, who passed away in 1952, did not eat for 40 years, no food, no water, and lived a contented life." (Source: none given.)

Food – Tea & Coffee

Tea and coffee are such vital substances that men and women devote much thought to the what, why, and how of making and handling a cup of the one or of the other. Four particular achievements stand out:

- How to Make a Cup of Tea, Officially
- It Takes Guts: Luak Coffee
- The Sociology of Canadian Donut Shops
- The Optimal Way to Dunk a Biscuit

How to Make a Cup of Tea, Officially

"We are delighted to have been recognised for what is the very important task of setting out the standards required to produce a proper cup of tea."

– Steve Tyler of the British Standards Institution,
in an interview with the *Guardian*

THE OFFICIAL CITATION

THE IG NOBEL LITERATURE PRIZE WAS AWARDED TO

The British Standards Institution for its six-page specification (BS 6008) of the proper way to make a cup of tea.

What is the proper way to make a cup of tea? The question has many answers, but only one of them is the official British Standard.

The Tea Standard was issued by the British Standards Institution, an organization known, as affectionately as a standards institution can be known, as "the BSI."

As with all standards promulgated by the BSI, the Tea Standard has a formal name, and it has a number. "Method for Preparation of a Liquor of Tea for Use in Sensory Tests" is standard number BS 6008.

BS 6008 has stood unchanged since 1980. In printed form it is six pages long, and it is valuable. The exact value is £20, the per-copy price at which it is sold by the British Standards Institution.

To those not steeped in the tea trade, the word "liquor" in the title may be confusing. The BSI points out that in this usage "liquor has no attachments to alcohol or spirits," instead meaning "a solution prepared by extraction of soluble substances."

What does it mean to make a cup of tea? Officially, it means to take "extraction of soluble substances in dried tea leaf, contained in a porcelain or earthenware pot, by means of freshly boiling water, pouring of the liquor into a white porcelain or earthenware bowl." The pot must have "its edge partly serrated and provided with a lid, the skirt of which fits loosely inside the pot."

BS 6008 is flexible. It includes provisions for making tea with milk ("pour milk free from any off-flavour into the bowl") or without.

Here is a much-abridged version of the British Tea Standard, BS 6008:

- Use 2g of tea – plus or minus 2% – for every 100ml of water.
- Tea flavour and appearance will be affected by the hardness of the water used.
- Fill the pot to within 4–6mm of the brim with freshly boiling water.
- After the lid has been placed on top, leave the pot to brew for precisely six minutes.
- Add milk at a ratio of 1.75ml of milk for every 100 ml of tea.
- Lift the pot with the lid in place, then "pour tea through the infused leaves into the cup".
- Pour in tea on top of milk to prevent scalding the milk. If you pour your milk in last, the best results are with a liquor temperature of 65–80 degrees C.

Altogether the British Standards Institution publishes more than 15,000 standards covering seemingly all aspects of commercial and daily life. Numerically, the Tea Standard comes just after BS 6007 ("Rubber-Insulated Cables for Electric Power and Lighting") and right before BS 6009 ("Hypodermic needles for single use: Colour coding for identification"). Other favorites from the 6000 series include:

BS 6094 – "Methods for Laboratory Beating of Pulp"
BS 6102 – "Screw Threads Used to Assemble Head Fittings on Bicycle Forks"
BS 6271 – "Miniature Hacksaw Blades"
BS 6310 – "Occluded-Ear Simulator for the Measurement of Earphones Coupled to the Ear by Ear Inserts"
BS 6366 – "Studs for Rugby Football Boots"
BS 6386 – "Tool Chucks with Clamp Screws for Flatted Parallel Shanks: Dimensions of Chuck Nose"

To many literary critics, these other standards do not approach the asthetic heights reached by BS 6008, "Method for Preparation of a Liquor of Tea for Use in Sensory Tests." The hot, steamy prose of BS 6008 epitomizes literature, propriety, and tea time, setting a standard to which all can sippingly adhere.

For their six-page classic, the British Standards Institution was awarded the 1999 Ig Nobel Prize in the field of Literature.

Reginald Blake of the British Standards Institution, devisers of the official British Standard for how to make a cup of tea. Photo: Sheila Gibson.

A representative of the British Standards Institution traveled, at the BSI's expense, from England to the Ig Nobel Prize Ceremony. Wearing a dark business suit, a top hat surmounted with a small teapot, and teacups dangling over his ears, Reginald Blake arrived at the airport in Boston and took the short cab ride to the

IG NOBEL PRIZES

PROFESSOR LIPSCOMB MAKES A CUP OF TEA

Among the Nobel Laureates participating in the 1999 Ig Nobel Prize Ceremony, one was so moved by the British Tea Standard that he presented BSI representative Reginald Blake with an additional, personal tribute. For the benefit of Mr Blake and the suddenly tea-crazed 1,200 audience members, William Lipscomb, the 1976 Nobel Chemistry Prize winner, showed and narrated a slide show called *Professor Lipscomb Makes a Cup of Tea.* The two images reproduced here give a glimpse of how the Harvard chemistry professor scientifically prepares his favorite beverage.

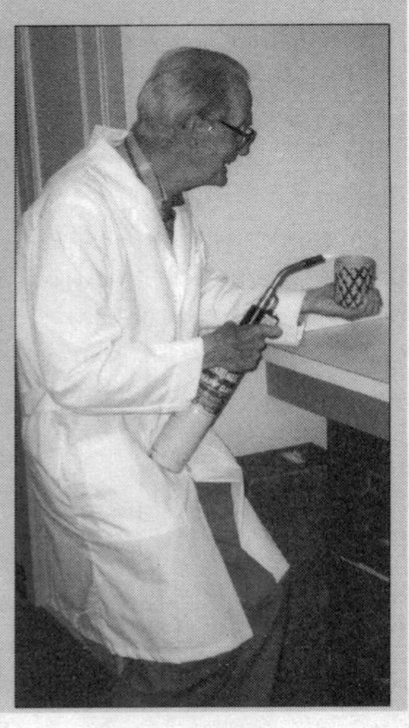

Highlights from the slide show *Professor Lipscomb Makes a Cup of Tea*, Nobel Laureate William Lipscomb's personal tribute to the winner of the 1999 Ig Nobel Literature Prize. Left: Professor Lipscomb obtains water. Right: Professor Lipscomb puts it to the boil.

ceremony at Harvard University. With every fiber of his very proper being, Mr Blake strained to make clear that he was representing not just his employer, but also the entire British tradition of making and drinking tea. He said:

"It has only taken us 5,000 years to develop a standard on how to make hot tea, so don't expect to see a cold-tea or an iced-tea one until the year 7000. By the way, we Brits have concluded that the Boston Tea Party was simply the first attempt to make iced tea on a grand scale. How do you make a cup of tea very quickly: you put 2g mass of tea per 100ml of water. I'm not going to do the conversion for you. Fill the pots within 4–6mm of the brim, and put on the lid. Brew for six minutes. Pour 5ml of milk into a cup and pour in the tea. In the best traditions, I'd like to thank BSI, the Boston Philharmonic, and Julius Caesar of 'I came, I saw, I had a cup of tea' fame. Thank you."

The audience pelted Mr Blake with paper airplanes, tea bags, and affection.

It Takes Guts: Luak Coffee

"Guests of an East Java plantation will almost certainly be offered kopi luak for breakfast. The secret of this delicious blend of coffee, usually explained only after the guest has drained his mug to the last drop, lies in the bean selection, which is performed by a luak, a species of civet cat endemic to Java. The luak will eat only the choicest, most perfectly matured beans which it then excretes, partially digested, a few hours later. Plantation workers then retrieve the beans from the ground, ready for immediate roasting."

– Indonesia Tourism Promotion Board

THE OFFICIAL CITATION

THE IG NOBEL NUTRITION PRIZE WAS AWARDED TO

John Martinez of J. Martinez & Company in Atlanta, for luak coffee, the world's most expensive coffee, which is made from coffee beans ingested and excreted by the luak (a.k.a. the palm civet), a bobcat-like animal native to Indonesia.

Luak coffee can be obtained directly from the luak in the Indonesian islands of Sumatra, Java and Sulawesi, or indirectly from J. Martinez & Company, Atlanta, GA, USA. Telephone 404-231-5465.

It takes guts to make luak coffee, and it takes guts, at least initially, to drink it.

The most expensive coffee in the world comes from Indonesia, where it is gathered by an animal, swallowed by an animal, deposited by an animal, collected by local human beings, and then (this is where the expense creeps in) shipped for mark-up and sale to giddy coffee fanciers in distant lands. A highly educated gentleman named John Martinez, born and bred to the coffee trade, is more responsible than any other individual for bringing this provocative yin-and-yang of a delight to the notice of a coffee-mad international community.

The luak, known formally as the palm civet, and more

Five young women serve up a surprise for the five Nobel Laureates – steaming hot mugs of luak coffee.

formally still by its scientific appelation Paradoxurus hermaphroditus, is a small, dark brown relative of the bobcat, dwelling in certain rainforest regions of Indonesia, and spending much of its time in trees. The luak is frugiferous, feeding on berries and pulpy fruits. It is believed to choose these when they are at the peak of perfect ripeness. The seeds pass through the animal's digestive system, emerging in pristine condition. What excites people about the luak is that it finds and eats the reddest, ripest (and only the reddest, ripest) coffee beans, passes them undigested through its entire system, and thus acts as a coffee connoisseur's connoisseur.

Luak coffee comes from the islands of Sumatra, Java and Sulawesi. Every year somewhere between 80 and 500 pounds (reliable figures are hard to come by in the luak coffee trade) of it are exported, mostly to Japan and the United States.

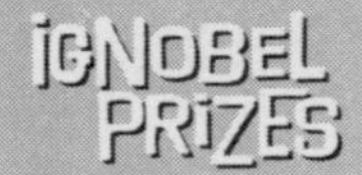

WHEN SCIENTISTS DRINK COFFEE

These photographs document the modern equivalent of Frederick the Great's famous public tasting of a potato. In this case it was five Nobel Laureates (one not visible in this shot) taking their first sips, in public or anywhere else, of luak coffee.

Nobel Laureates Dudley Herschbach (Chemistry '86), Sheldon Glashow (Physics '79), Joseph Murray (Medicine '90), and William Lipscomb (Chemistry '76) take their first taste of luak coffee. Richard Roberts (Physiology or Medicine '93) is not visible in this photo. Photo: Jeff Pietrantoni/*Annals of Improbable Research*.

Professor Lipscomb performs an analysis of the beverage. Photo: Jon Chase/ Harvard News Office.

The animal does, in fact, pre-process the beans, but not exactly in the way laypersons might expect. Coffee beans grow wrapped in a sticky outer protein covering that has to be removed before the beans are roasted. These days, most coffee beans are de-pulped with machines, but luak coffee beans are de-pulped the all-natural way – using just the luak's intestinal fortitude. The luak deposits the beans without pulp and, say those who know, generally without much added packaging material. Harvesting the beans is simply a matter of walking about and gathering up the little piles left by the luak.

As one commentator has put it: "the unusual taste and aroma of the coffee is usually attributed to the fact that it has been partially digested by the animal before it's harvested." The animals do have well-developed anal scent glands, but luak coffee enthusiasts insist that these add little discernable taste to the coffee beans. Almost all agree that luak coffee is rich and full-bodied. While some describe the taste as "gamey," many insist it is the most delicious of all brews.

John Martinez grew up in a coffee-growing family in Jamaica. He moved to the United States and founded J. Martinez & Company, establishing a reputation as one of the world's great coffee experts and coffee merchants. It is largely through his efforts that luak coffee rose from utter obscurity to the level of renown it enjoys today. Martinez took the product on, he said in an interview with the *Atlanta Journal-Constitution*, because "when you are dealing at the top of the market in the world's second most-traded commodity, it is necessary to look for something that is unique."

Why is luak coffee not just pricey, but supremely expensive? Apparently to ensure that it makes such a memorable statement. Anyone in proximity to a luak and some coffee trees, though, can, of course, pick some up at no monetary cost.

For bringing this rare, and possibly delicate, taste sensation to the world stage, John Martinez won the 1995 Ig Nobel Prize in the field of Nutrition.

The winner journeyed from Atlanta, Georgia to the Ig Nobel Prize Ceremony, at his own expense. Wearing an expensively and beautifully tailored business suit, and standing with aplomb and dignity, he accepted the prize with an intensely personal tribute to the well-traveled coffee:

"In keeping with the spirit of the evening, I wrote a little poem. It's called *Ode to the Luak*.

"Luak, luak-,"
The cry is heard.
Is it a plane? Is it a bird?
Alas, my friends, it's not like that.
My star's a first cousin to a pole cat.

Luak, luak-
We ask, "What is it?"
Britannica says it's a palm civet,
Genus Paradoxurus, species hermaphroditus,
-Scientific information to excite us.

Luak, luak-
Where do you dwell?
On the isle of Sumatra, where it's hot as hell,
Who under the cover of darkest night
Hunts for ripe coffee cherries red and bright.

Luak, luak-
After you've gorged,
A new taste sensation is forged.
For all gathered here, this is the scoop –
We're drinking coffee made from your poop.

As Martinez finished his acceptance speech and dabbed his dampened eyes, five young dancers sashayed purposefully into the hall. Each was wearing a lab coat, and each was carrying a mug of fresh-brewed, steaming luak coffee. These were served to the five Nobel Laureates.

The Laureates sat staring alternately down at their coffee mugs and up at each other. It was a moment when everyone in the room seemed to know exactly the thought that inhabited each of the great men's minds: "If one of us drinks this . . . then we *all* have to drink it." After a long, giggly, anguished pause, and several halting attempts to raise coffee mugs to mouths, first one Nobel Laureate, then all of them gently sipped the rare treat, and then drank deep. And at that point these five revered, dignified public figures all began making poop jokes. When it became clear that they would not stop of their own accord, the master of ceremonies wearily shamed them with a muttered "All right, boys ..."

In the spirit of fairness, the master of ceremonies then borrowed a cup, took a hearty swig of luak coffee, and swallowed. "How's it taste?" yelled a heckler. "Even better than you might expect," came the reply.

The Sociology of Canadian Donut Shops

"Donuts are a mass-production and mass-consumption product, [yet] donuts are also a quirky receptacle for the politics of identity in Canada."

–from Steve Penfold's treatise "The Social Life of Donuts . . ."

THE OFFICIAL CITATION

THE IG NOBEL SOCIOLOGY PRIZE WAS AWARDED TO

Steve Penfold, of York University in Toronto, for doing his PhD thesis on the sociology of Canadian donut shops.

His PhD thesis, prepared for the History Department at York Univerity, is titled "The Social Life of Donuts: Commodity and Community in the Golden Horseshoe, 1950–1999." A preliminary essay version was published as "'Eddie Shack was no Tim Horton': Donuts and the Folklore of Mass Culture in Canada," in the book *Food Nations: Selling Taste in Consumer Societies*, Warren Belasco and Philip Scranton, editors, Routledge, 2001, pp. 48–66.

As a graduate student in the history department of York University, Steve Penfold had to choose a topic that required original research on a subject of interest to scholars in his field. Penfold chose to write about the place of donut shops in the social fabric of Canada.

Canadians eat more donuts per capita than any other people on earth. The largest share of those donuts is obtained from and/or eaten at the almost 2,000 Tim Horton's donut shops spread across Canada. These shops, named after a professional hockey player now deceased, outnumber McDonald's restaurants. For many Canadians, and for many Canadian towns, the Tim Horton's, with its donuts, its coffee, and its well-heated place of shelter from the winter that dominates the Canadian psyche, is the key to social life. Penfold explained to the *Wall Street Journal* that "In England, people go to the local pub to socialize; in Canada, they go to the local donut shop."

Here are some extracts from his work.

"In Canada, the donut is widely believed to be the unofficial national food. Indeed, the fatty treat is celebrated as a sort of ironic replacement for the dramatic national symbols found south of the 49th parallel. We consume American products, yet somehow crave a more 'genuine' Canadian mass culture experience, like a Tim Horton's coffee on a February morning. Expatriate Canadians speak of associating a trip to the donut shop with returning home.

"The rise of these national associations seems curiously disconnected from the origins and the fate of the commodity itself. As with much of 20th-century Canadian economic history, big-time donut retailers developed in Canada as branch plants or Canadian-owned versions of US mass-production ideas. In 1995, Tim Horton's was sold to Wendy's. While Canadians are normally sensitive to the threat of American-owned companies, the sale of this 'national institution' to an American hamburger company did not seem to affect Tim Horton's link to national mythology.

"If we believe that the dynamic of mass culture is to degrade production on the one hand and to reduce social experience to consumption on the other, then the donut takes on considerable analytic power. In Canada, the donut is mainly produced by large companies, sold in cookie-cutter shops across the country, and served by low-wage workers doing carefully defined, unskilled jobs. Yet the donut is also a vehicle for ironic depictions of Canadian life. Ultimately, the effect of donut folklore – the nature of its mediation of structure and identity, of mass and community – remains ambiguous."

For finding new meaning in donuts, Steve Penfold won

the 1999 Ig Nobel Prize in the field of Sociology.

The winner traveled to the Ig Nobel Prize Ceremony, at his own expense. In accepting the Prize, he said:

"Well, I understand that in America, doing a PhD thesis on the sociology and the history of donut shops might be considered a little off. I don't know. In Canada, this is considered quite normal. In fact, quite noble, if I may use that word. In fact, people come up to me in the street and say, "Way to go, buddy. Good job, eh." Because you may not realize the donut is Canada's national food. Poems are written in tribute. Songs sing the praises of our fatty treat. Yet, as a Canadian, I am proud to say that to get this award in the city that originated the chain donut store, Boston, Massachusetts, indeed, you are so well supplied with donuts, I would move here if it weren't for only one problem: as I walk around, there's too many Americans! A whole city filled with Americans. And I have no problem with that, you being Americans in the privacy of your own home, but don't do it on the street where everybody else can see."

Writing a doctoral thesis on the sociology of Canadian donut shops is not the least stressful of all possible experiences. In a departmental newsletter in 2001, one of Steve Penfold's fellow graduate students touched on this:

"I went to a session on tips and strategies for thesis writers. Did I learn anything? Well, yes. Steve Penfold, probably unwittingly, taught me a lesson. On that day, I heard Steve – the donut man, for those of you don't know him by his real name – ask twice how not to kill everybody around him while writing his thesis. During the session, when the question came up again, I could not help but to worry about his state of mind which obviously originated from the fact that he has been working on his thesis... for too long? Steve's aggressiveness plunged me into deep thoughts about my work and mostly about my life outside academia ... I remain hopeful that Steve Penfold will only stay famous for his passion for donuts and not because of some horrific event involving the killing of his family."

The Optimal Way to Dunk a Biscuit

"Here is the equation for dunking in hot tea:"

$$L^2 = \frac{\gamma \times D \times t}{4 \times \eta}$$

distance that tea penetrates (squared) = surface tension of tea x average pore diameter in biscuit x time / 4 x viscosity of tea

–Len Fisher, in an unpublished, but widely distributed, paper elucidating the physics of biscuit dunking

THE OFFICIAL CITATION

THE IG NOBEL PHYSICS PRIZE WAS AWARDED TO

Dr Len Fisher of Bath, England, and Sydney, Australia, for calculating the optimal way to dunk a biscuit

... and ...

Professor Jean-Marc Vanden-Broeck of the University of East Anglia, England, and Belgium, for calculating how to make a teapot spout that does not drip.

To dunk a biscuit, to pour tea from a pot – these are activities of joy and grace, exemplars of the art of living. And, like everything else in the universe, they are ripe for scientific analysis. Two men of great learning and tenacity, working independently and thirsting for knowledge and tea, drew out nature's secret recipes for how to dunk and how to pour.

Len Fisher was the first person to recognize, or at least the first person to bother to recognize, that dunking a biscuit makes more sense if you use a belt-sander, an X-ray machine, scales, and a microscope, and apply the Washburn equation for capillary flow in a porous material, and if you are a physicist, which he is.

Happily for the biscuit-dunking public, Dr Fisher took

the time (aided by research funding from McVitie's) to translate the technicalities so they would be palatable to the average dip-and-munch Joe in the street. He produced a simpler version of the equation, and also some guidelines:

• Different biscuits have different optimal dunking times (a Gingernut biscuit reaches its dunkability zenith at approximately three seconds, a digestive biscuit at eight seconds).
• Biscuits with chocolate on one side are optimally dipped biscuit-side down; or failing that, at an angle.

Being a scientist, Dr Fisher also produced a nice set of numbers and charts.

Meanwhile, in Norwich, a mathematician named Jean-Marc Vanden-Broeck was reaching the climax of a 17-year-long effort to design, at least theoretically, a non-dripping teapot spout. A professor at the University of East Anglia, the Belgian native is an expert in the difficult field of fluid dynamics.

Why the magnificent obsession with teapots? Partly it's because this really does constitute a very interesting and difficult problem in the physico-mathematics of fluids and surfaces. Though it gets little public acclaim, there is a rich history of scientific inquiry into how things drip. [The interested reader is urged to delve into the prestigious physics research journal *Physical Review Letters*, which in recent years has published papers titled "Theoretical Analysis of a Dripping Faucet" (2000), "Suppression of Dripping From a Ceiling" (2001), and the provocative "Dripping Faucet With Ants" (1998).]

But there may be another reason. Mathematicians – good ones, anyway – are much in demand to give talks at universities around the world. Almost inevitably, an academic talk is preceded by tea. After getting dripped upon one too many times, a scientist with pertinent expertise and experience could very well let annoyance become a spur to his keen imagination.

Especially after he discovered the secret to a dripless teapot, Professor Vanden-Broeck was being invited to figure in events such as this one advertised for 1 June, 2001, at the University of Edinburgh:

> SEMINAR
>
> JEAN-MARC VANDEN-BROECK (UEA)
>
> Seminars Take Place at 3.30 p.m. Everybody is Welcome.
> TEA IS SERVED AT 3 p.m. IN THE COMMON ROOM.

Similarly, Dr Len Fisher, discoverer of the optimal way to dunk a biscuit, became a popular figure on the scientific tea-and-biscuits scene.

For their scientific discoveries with teapots and biscuits, Jean-Marc Vanden-Broeck and Len Fisher shared the 1999 Ig Nobel Prize in the field of Physics.

Professor Vanden-Broeck could not, or would not, attend the Ig Nobel Prize Ceremony, but Dr Fisher traveled from Bristol, England, to Harvard University at his own expense, to accept his Prize.

There he said: "I thank you. 200 years and finally we have a British winner at a Boston tea party. In honor of this splendid audience, and with the assistance of at least one Nobel Laureate who does not yet know that he is a volunteer, I wish to introduce an extension of my theory. The extension concerns the very difficult and abstruse problem of doughnuts. Doughnuts. Kindly take the doughnut. And now, Professor Glashow, I wish you to demonstrate to this uncomprehending audience my new method of dunking a doughnut."

At which point, Dr Fisher held up a miniature basketball

net, and Nobel Laureate Sheldon Glashow dunked a donut through it. Moments later, in a tribute to the accomplishments of tea, donuts, and science, a giant artificial donut descended on a rope and pulley from the back ceiling of Sanders Theatre, flew over the heads of the audience, and reaching the stage, dunked itself into an attractive, leggy, tap-dancing teacup.

Several days later, Dr Len Fisher returned to his laboratory, bent on extending his work. He emerged, a year later, with a pair of remarkable discoveries. First, that a biscuit yields more flavor if dunked in a milky drink rather than in plain tea; and, second, that a biscuit dunked in lemonade is not a good thing.

Education

All sorts of things can teach you a lesson. This chapter describes four great achievements in the field of education:

- Therapeutic Touch
- Banning the Beaker
- Deepak Chopra
- Dan Quayle

Therapeutic Touch

"There is much that I do not understand about this strange occurrence."

–Dolores Krieger, in her book *The Therapeutic Touch*

THE OFFICIAL CITATION

THE IG NOBEL SCIENCE EDUCATION PRIZE WAS AWARDED TO

Dolores Krieger, Professor Emerita, New York University, for demonstrating the merits of Therapeutic Touch, a method by which nurses manipulate the energy fields of ailing patients by carefully avoiding physical contact with those patients.

Dolores Krieger has a number of publications about Therapeutic Touch. For a good overview see her book *The Therapeutic Touch: How to Use Your Hands to Help or to Heal*, Prentice Hall, 1979. A set of audio cassettes of the author reading her book is available from the Ventana Catalog in Ontario, California, as is a videotape. Emily Rosa's study testing the theory behind Therapeutic Touch was published as "A Close Look at Therapeutic Touch," Linda Rosa, Emily Rosa, Larry Sarner, and Stephen Barrett, *Journal of the American Medical Association*, vol. 279, April 1, 1998, pp. 1005–10.

Talented healers (and nurturing mothers) always have known the power of physically touching another person, or of even sitting nearby and making one's presence known. Touching, talking, showing interest – any and all of these make people relax and feel cared for. It's good for whatever ails a body.

This was all old hat until Dolores Kreiger took the power of Touch, took away the "touch" part, and added to it to the power of Science.

Dolores Krieger is a professor (now retired) of nursing at New York University. In her book, Professor Krieger explained that she learned about hand waving from a man named Colonel Oskar Estabaney. Colonel Estabaney, she says, rose to fame in the Hungarian cavalry, where he healed pet animals by praying over them. When she knew him, Colonel Estabeney was "a well-built man with cheery blue eyes and a frequent smile." Krieger says she often watched him magnetize rolls of cotton batting:

"He would distribute the magnetized cotton to the healees after having it near his person during the night; some of these patients have told me that, even after a year, they could still feel an energy flowing through the cotton."

That may be the first report, anywhere, ever, about magnetic cotton.

Professor Krieger's book mentions lots of huge, breakthrough scientific discoveries like that one – discoveries so big, so surprising that each of them could win her a Nobel Prize. Professor Krieger modestly doesn't even suggest that there is anything remarkable about them.

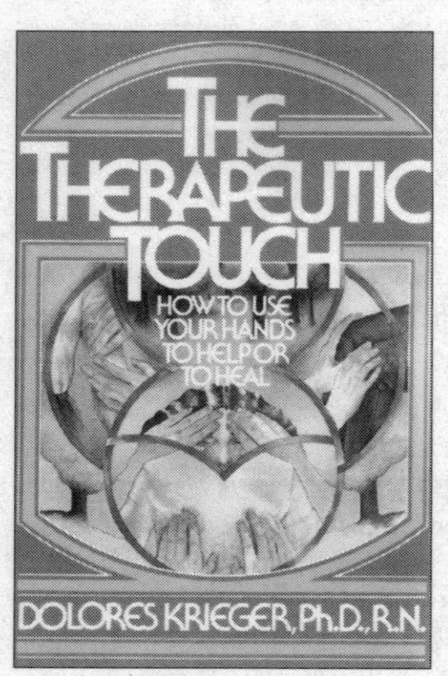

One of Dolores Krieger's books explaining the technical basis of Therapeutic Touch.

Another example is her discovery, with Colonel Estabaney, that when she passes her hand near sick people, "a significant change occurs in the hemoglobin component of their red blood cells."

She and Colonel Estabaney also discovered that people's internal organs have special energy fields that are unknown to science. Professor Krieger can "sense" these fields from a distance, and manipulate them, too. She named this "Therapeutic Touch."

She became better at Therapeutic Touch by talking with psychics and by reading books about spirit healing. Eventually she became a professor of nursing at New York University. There she began teaching student nurses about Therapeutic Touch. She also persuaded other nursing schools to teach their students about Therapeutic Touch.

Her 1979 book, titled *The Therapeutic Touch*, explains basic topics such as how to place a hand over a cotton pad without touching it (page 28), and how to make a dousing rod with a wire clothes hanger (page 31).

The book also covers advanced topics: how to use a stained-glass window to calibrate one's conceptualization of a modulation of energy (page 60); how to transfer an energy field to a dog (page 118); how to energize cotton (page 120); how to perform Therapeutic Touch on oneself (page 144); and how to use Therapeutic Touch to fix a broken electronic fetal monitor machine (page 146).

With the publication of her book, Professor Krieger says, Therapeutic Touch became an important medical technique based on solid, scientific theory.

The nice thing, or, in some cases, unfortunate thing, about a scientific theory is that when people learn about it, they want to test it, to see whether it holds water. If a theory stands up to good, stiff tests, that speaks well for it. If fails simple tests, that says something, too.

In the town of Loveland, Colorado, nine-year-old Emily Rosa was looking for a project to do for her school science fair. Emily's mother, a nurse, told her about Therapeutic Touch. Emily decided that the scientific theory of Therapeutic Touch was really neat, and really interesting. She came up with a simple way to test whether one person really can sense another person's energy fields. Emily and her mom recruited some skilled Therapeutic Touch practitioners, and had them stick their arms through holes in a cardboard screen. Emily would then ask them to say when they could sense Emily's hand near theirs. It turned out they couldn't.

A doctor heard about Emily's project and suggested that Emily and her parents team up with him to write a formal report and submit it to a respected medical journal. They did, the journal published the report (on April Fools' Day 1998), and two things happened. First, Emily became a minor celebrity – the youngest person ever to have published a report in the prestigious *Journal of the American Medical Association*. Second, the public for the first time heard the phrase "Therapeutic Touch." Some were delighted, many were not – especially when they learned how many hospitals had been charging them fees, sometimes substantial, to have nurses not-quite-touch them.

Through it all, Dolores Krieger persevered.

For inspiring so many people to think about science, Dolores Krieger won the 1998 Ig Nobel Prize in the field of Science Education.

The winner could not, or would not, attend the Ig Nobel Prize Ceremony, and so the Ig Nobel Board of Governors arranged for Emily Rosa to come and deliver that year's Keynote Speech. Emily, who at the time was 11 years old, traveled to Harvard with her parents. A remarkably poised and thoughtful child, she had this to say:

"Therapeutic Touch nurses say they can cure people of anything by using their hands to feel and manipulate something they call the 'human energy field.' Dolores Krieger, the inventor of Therapeutic Touch, even says the dead have been brought back to life – that all you have to do is send

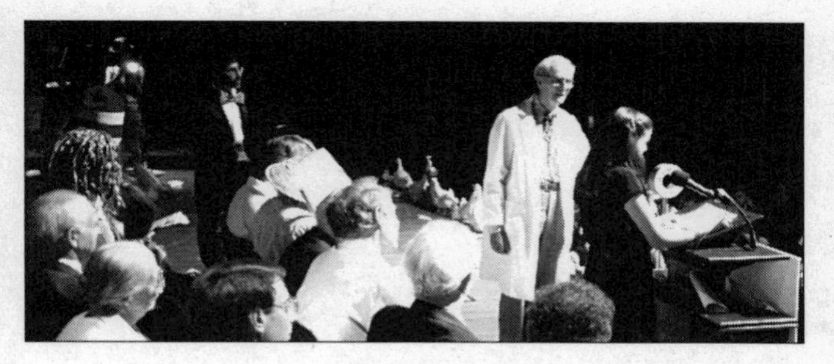

11-year-old Emily Rosa delivers the Keynote Address at the 1998 Ig Nobel Prize Ceremony. For a school science fair project when she was nine, Emily tested the scientific basis of Therapeutic Touch. Nobel Laureate William Lipscomb stands behind Emily. Photo: Eric Workman/ *Annals of Improbable Research*.

this energy to cells where it starts 'protein pendulums' swinging again. Professor Krieger also uses her energy field to talk to trees. I don't understand what she means by that.

"TT nurses say the human energy field stretches out forever – they say that they can now treat patients over the phone. They even publish ads that call this stuff 'healing on the cellular level.'

"But I wanted to see for myself if they can even find one of these 'human energy fields.' I learned with my experiment that they can't.

"I also learned a lot about my own personal energy field. Therapeutic Touch nurses say that my personal energy field interfered with the test, because it is too healthy to feel, or it's too big, or it's too little, or that it's turned off because I'm a skeptic. Others say it gets blown away by the air conditioning or shoots out wildly in all directions because I'm becoming a teenager.

"Watch out if you get sick. Some nurses still do Therapeutic Touch and other weird things."

Banning the Beaker

"In this section, 'chemical glassware laboratory apparatus' means any equipment designed, made, or adapted to manufacture a controlled substance, including:

1) condensers
2) distilling apparatus
3) vacuum dryers
4) three-neck flasks
5) distilling flasks
6) ..."

–from the Texas glassware law

THE OFFICIAL CITATION

THE IG NOBEL CHEMISTRY PRIZE WAS AWARDED TO

Texas State Senator Bob Glasgow, wise writer of logical legislation, for sponsoring the 1989 drug control law which makes it illegal to purchase beakers, flasks, test tubes, or other laboratory glassware without a permit.

The glassware regulations are part of the "Texas Controlled Substances Act" as published in article 4476–15 of Vernon's Texas Civil Statutes and Sections 17–20 of the Texas Health and Safety Code (481.080) as amended by the 71st Legislature, Regular Session, 1989. For the administrative interpretation of how the law is carried out, see the Texas Administrative Code, specifically 37–TAC–13.131.

Some politicians care only about protecting their jobs, but others try very hard to protect the public. A determined few keep watch for things that might be dangerous and can be banned. Texas State Senator Bob Glasgow was extremely determined and extraordinarily watchful.

In 1989, Bob Glasgow persuaded his colleagues in the Texas State Legislature that laboratory glassware – beakers, flasks, and the like – should be considered illegal drug paraphernalia. The legislature agreed. Under Texas law, it is now a Class A misdemeanor to buy, sell or even give these items without a permit from the state. Those who flout the law are subject to a jail term of up to a year and a fine of up to $4,000.

The Texas Department of Public Safety handles the permit applications. There is an eight-page set of instructions. The permit itself is seven pages long.

(1) Apparatus—Any chemical laboratory equipment designed, made, or adapted to manufacture a controlled substance or a controlled substance analogue including:

(A) the following items listed under the Health and Safety Code, Chapter 481, §481.080(a):

(i) condensers;

(ii) distilling apparatus;

(iii) vacuum dryers;

(iv) three-neck flasks;

(v) distilling flasks;

(vi) tableting machines;

(vii) encapsulating machines; and

(B) the following additional items determined by the director to jeopardize public health and welfare by evidenced use in the illicit manufacturer of controlled substances or controlled substance analogues:

(i) filter funnels, buchner funnels, and separatory funnels;

(ii) erlenmeyer flasks, two-neck flasks, single-neck flasks, round bottom flasks, florence flasks, thermometer flasks, and filtering flasks;

(iii) soxhlet extractors;

(iv) transformers;

(v) flask heaters;

A portion of the Texas glassware law.

Anyone outside Texas should now know that it is illegal to send any of the old laboratory favorites – an Erlanmeyer flask, a Florence flask, a glass funnel, or even the exotically named Soxhlet extractor – to somebody in Texas without first obtaining a permit to do so. The state of Texas will pursue those who recklessly flout Bob Glasgow's restrictions.

Bob Glasgow left the State Senate in 1993. He is now a lawyer in private practice in Stephenville, Texas. His firm's web site (www.robertjglasgow.com) proudly notes that *Texas Monthly* magazine included Glasgow on its "10 Best Legislators" list in 1987. The web site does not mention that in 1989 the magazine moved Glasgow to its "10 Worst Legislators" list. Bob Glasgow's web site does, though, mention, without explanation, the curious fact that he served as Governor of Texas on May 11, 1991.

For protecting the public from test tubes and beakers, Bob Glasgow won the 1994 Ig Nobel Prize in the field of Chemistry.

The winner could not, or would not, attend the Ig Nobel Prize Ceremony, and so the Ig Nobel Board of Governors arranged for a major manufacturer of laboratory glassware to give a tribute. Tim Mitchell of Corning, Inc came to the Ceremony and presented the following thoughts:

"I am here to accept this in lieu of the actual winner. Tonight I'll use this forum to make a few comments on a hot social and scientific issue brought to light by the law makers in Texas. That topic is the unregulated and unrestricted sale of test tubes, beakers, and other laboratory apparatus in America.

"There is a grass-roots movement out there to convince the state of Texas to amend their laboratory glassware law. Instead of outlawing glassware altogether, this group would like to see a five-day cooling-off period. They feel this will be enough to discourage people from purchasing a beaker and then using it in a fit of rage to harm themselves or others.

"Part of me wonders, will a waiting period be enough? You see, it only starts with a test tube. You think to yourself, 'Hey, it's only a test tube, for God's sakes.' Pretty soon, though, the rush from a test tube isn't enough. You want to experiment more and more. Then before you know it, you're laying in the corner of a lab somewhere with a Soxhlet extractor apparatus in one hand, a three-neck flask in the other, strung out and begging for grant money."

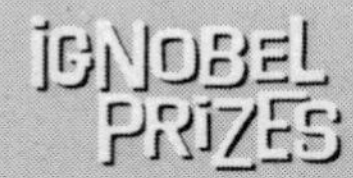

OBEY THE LAW!

If you live in Texas, or are planning to visit Texas, the Ig Nobel Board of Governors urges you to forswear the use of illegal flasks, beakers, and test tubes. But if you feel you must purchase any of these items, please do it legally. The Texas Department of Public Safety's official permit application form is available on their web site. Please use it.

Deepak Chopra

"Biology will have to change now, and medicine with it. Contrary to what physicians currently suppose, the abnormal pancreas of a diabetic is not as real as the distorted memory that has wrapped itself inside the pancreatic cells. This realization opens the door for quantum healing."

– Deepak Chopra, in his book *Quantum Healing* . . .

THE OFFICIAL CITATION

THE IG NOBEL PHYSICS PRIZE WAS AWARDED TO

Deepak Chopra of the Chopra Center for Well Being, La Jolla, California, for his unique interpretation of quantum physics as it applies to life, liberty, and the pursuit of economic happiness.

Deepak Chopra has published numerous studies about quantum issues. Two of the best-known are *Quantum Healing: Exploring the Frontiers of Mind/Body Medicine, and Ageless Body, Timeless Mind: The Quantum Alternative to Growing Old.*

For more than a century, the juiciest mystery in physics has been about quantum mechanics. Specifically: why do the teeny, tiny little quanta of energy and matter seem to behave so weirdly? Physicists assume that if they work hard enough to understand the weirdness, it will turn out to be not weird at all. But one lone, strong voice sings a different tune. To Deepak Chopra, MD, the weirdness itself is the important thing – and it's something to celebrate, not to understand.

Quantum. The idea started in the year 1900, when Max Karl Ernst Ludwig Planck realized that energy seems to come in extremely tiny amounts – which he called "quanta," all of which are the same size and none of which can be divided into anything smaller. Scientists soon discovered that Planck was correct. He was awarded a Nobel Prize in Physics, and nearly all the Nobel Physics Prizes

One of Deepak Chopra's many best-selling books about the quantum nature of nature's quantum nature.

awarded ever since have, one way or another, been for showing that the weirdness is a little less weird than it at first seemed.

Quantum. Deepak Chopra's web site (www.chopra.-com) says that: "He is widely credited with melding modern theories of quantum physics with the timeless wisdom of ancient cultures."

Quantum. In 1989, Deepak Chopra published a book called *Quantum Healing: Exploring the Frontiers of Mind/Body Medicine*. Physicists who see the book typically say that it uses the word "quantum" in ways that would never occur to them. A curious thing about this book called *Quantum Healing* is how seldom the word "quantum" appears in it. Most of those appearances are in the brief chapter called "The Quantum Mechanical Human Body," which contains the following passage:

"The discovery of the quantum realm opened a way to follow the influence of the sun, moon, and sea deeper into ourselves. I am asking you there only in the hope that there is even more healing to be found there. We already know that a human fetus develops by remembering and imitating the shapes of fish, amphibians, and early mammals. Quantum discoveries enable us to go into our very atoms and remember the early universe itself."

Quantum. "The human body first takes form as intense

but invisible vibrations, called quantum fluctuations, before it proceeds to coalesce into impulses of energy and particles of matter." (Source: Deepak Chopra)

Quantum. "The most important routine to follow is transcending: the act of getting in touch with the quantum level of yourself." (Source: Deepak Chopra)

Quantum. "Quantum health is based on the idea that we are always, forever, in transition." (Source: Deepak Chopra)

Quantum. The universe consists of a "field of all possibilities" called "the field of pure potentiality", and also the "quantum soup." (Source: Deepak Chopra)

Quantum. At the Chopra Center For Well Being, in La Jolla, California, one can dine at the Quantum Soup Cafe.

Quantum. Inspired by Deepak Chopra, the American Academy of Quantum Medicine, based in Iselin, New Jersey, provides Certification in Quantum Nutrition for health care providers in the fields of massage therapy, acupuncture, and nutritional counseling, and for certified nutritionists and registered nurses and dieticians, and also for "health care professionals with a doctorate degree including: MD, OD, DC, DDS, PhD, ND, OMD and IMD."

Quantum. Inspired by Deepak Chopra, a Dr Stephen Wolinsky developed the fields of Quantum Psychology and Quantum Psychotherapy. Dr Wolinsky also wrote a book called *Quantum Consciousness*, which is said to bring us full circle in understanding the reality of our inner child.

Quantum. The concept that won Deepak Chopra the 1998 Ig Nobel Prize in the field of Physics.

The winner could not, or would not, attend the Ig Nobel Prize Ceremony. At the Ceremony, two distinguished Harvard physics professors paid tribute to him.

Roy Glauber, the Mallinckrodt Professor of Physics, and one of the youngest scientists to have worked on the original atomic bomb project at Los Alamos, said:

"There is not much that I need to tell you about relativistic quantum mechanics. There is not much I *can* tell you about relativistic quantum mechanics. Its achievements in the world of atoms and particles have been great. Its successes, on the other hand, in the world of psychiatry and emotional well-being have been few. And it has certainly not been known for them, particularly. Not, that is, until the recent work of tonight's honoree. Success, of course, is a matter of definition. Relativity and quantum mechanics applied to personal well-being and psychiatry may or may not have done good, but they have certainly done well."

Sheldon Glashow, Higgins Professor of Physics, and winner of a 1979 Nobel Prize in Physics, said:

"I am honored to be here to discuss one of our Laureates. I am one of the few people in this room who have met with him, dined with him, spoken with him. He is held up as a role model to young men and women – high school students throughout the world – at the American Academy of Achievement each year. He is indeed a self-made man.

"Who else could have imagined quantum nutrition? I stand in awe of this man and his accomplishments. I, too, like Professor Glauber, teach a course called 'Relativistic Quantum Mechanics.' Its lectures are prepared about as well as this lecture is prepared. But I have to say that my course has little more to do with relativistic quantum mechanics than Professor Deepak Chopra's enormous and wonderful opus.

"He is a deserving and wonderful Laureate. Let's hear it

for him. DEE-PAK! DEE-PAK! DEE-PAK!"

At this point, the entire crowd joined Professor Glashow in chanting.

Dan Quayle

"What a waste it is to lose one's mind. Or not to have a mind is being very wasteful. How true that is."

– US Vice President Dan Quayle, in a 1989 speech to the United Negro College Fund

THE OFFICIAL CITATION

THE IG NOBEL EDUCATION PRIZE WAS AWARDED TO

J. Danforth Quayle, consumer of time and occupier of space, for demonstrating, better than anyone else, the need for science education.

Dan Quayle was Vice President of the United States from 1989–1993. During that time, he was also Chairman of the National Space Council, and a self-declared champion of education. His was the most inspirational teaching voice of the 20th century. Dan Quayle's words caused people to wonder. He sparked a universal interest in rhetoric and in logic; he made people everywhere appreciate the value of learning.

He leaves a legacy of lessons that are difficult to understand, and even harder to forget. His words sum up the man perhaps better than any description can. Here are some of Dan Quayle's observations.

- If you give a person a fish, they'll fish for a day. But if you train a person to fish, they'll fish for a lifetime.
- Quite frankly, teachers are the only profession that teach our children.
- We are not ready for any unforeseen event that may or may not occur.

- I have made good judgments in the past. I have made good judgments in the future.
- One word sums up probably the responsibility of any vice president, and that one word is "to be prepared."
- I stand by all the misstatements that I've made.
- My friends, no matter how rough the road may be, we can and we will never, never surrender to what is right.
- [The book *Nicholas, and Alexandra*] was a very good book of Rasputin's involvement in that, which shows how people that are really very weird can get into sensitive positions and have a tremendous impact on history.
- I believe we are on an irreversible trend toward more freedom and democracy, but that could change.
- A low voter turnout is an indication of fewer people going to the polls.
- People are not homeless if they're sleeping in the streets of their own hometowns.
- We [the people of the United States] have a firm commitment to NATO, we are a part of NATO. We have a firm commitment to Europe. We are a part of Europe.
- The global importance of the Middle East is that it keeps the Far East and the Near East from encroaching on each other.
- Bank failures are caused by depositors who don't deposit enough money to cover losses due to mismanagement.
- It isn't pollution that's harming the environment. It's the impurities in our air and water that are doing it.

- [It's] time for the human race to enter the solar system.
- Mars is essentially in the same orbit... Mars is somewhat the same distance from the Sun, which is very important. We have seen pictures where there are canals, we believe, and water. If there is water, that means there is oxygen. If oxygen, that means we can breathe.
- Space is almost infinite. As a matter of fact, we think it is infinite.
- If we do not succeed, then we run the risk of failure.

For so often inspiring people to stop and think, United States Vice President J. Danforth Quayle was awarded the 1991 Ig Nobel Prize in the field of Education.

The winner could not, or would not, attend the Ig Nobel Prize Ceremony.

Literature

Literature is the way many people leave their mark in history. In the following seven cases, it is how people earned Ig Nobel Prizes:

- Farting as a Defence Against Unspeakable Dread
- The Apostrophe Protection Society
- 976 Co-Authors in Search of a Title
- 948 Titles in Search of an Author
- Chariots of the Gods
- The Father of Junk E-Mail
- Up Theirs

Farting as a Defence Against Unspeakable Dread

"This paper describes some features of the behaviour of a severely disturbed adopted latency boy ... When feeling endangered, Peter has developed a defensive olfactory container using his bodily smell and farts to envelop himself in a protective cloud of familiarity against the dread of falling apart, and to hold his personality together. In the paper, Fordham's views of development and Anzieu's concept of psychic envelopes constitute the theoretical underpinning. Bion's concepts of beta- and alpha-elements are discussed in relation to Jung's views on symbolic development and psychological containment."

–from Mara Sidoli's report

THE OFFICIAL CITATION

THE IG NOBEL LITERATURE PRIZE WAS AWARDED TO

Dr Mara Sidoli of Washington, DC, for her illuminating report, "Farting as a Defence Against Unspeakable Dread."

Their study was published in *Journal of Analytical Psychology*, vol. 41, no. 2, 1996, pp. 165–78.

One of the world's greatest Jungian child psychoanalysts took on one of the toughest, smelliest cases in memory. Three years later she had much to be proud of, and for posterity's sake put pen to paper.

Among her peers, Mara Sidoli was renowned for her willingness to handle cases that had worn out or otherwise defeated others. The case of Peter proved especially challenging, for Peter was a severely disturbed latency boy who smelled putrid.

"Latency" is the term Sigmund Freud used to describe the ages 7–12 – the only period in people's lives, Freud thought, when they are not obsessed with sex. Peter was obsessed with other things.

A local hospital referred this curious patient to her. From the start, she knew this would not be an easy case:

"Peter held loud conversations with imaginary beings and made loud anal farts as well as farting noises with his mouth whenever he became anxious or angry. He was encopretic [often defecated in his pants, despite having been toilet trained] under stress."

Sidoli recognized early on that "his anxiety made him act to test his parents' commitment to him ... Even though I liked him immediately, I recognized that our working together would bring some extremely difficult times."

Her assessment was correct.

After several weeks of slow progress, she made a comment Peter didn't like. The reaction was immediate, she reports. "He defended himself by jumping up and down, screaming, and farting in a state of confusion and panic."

More weeks of steady, if slight, progress brought the boy to a point where he no longer defecated in his pants. Therapist and patient began meeting more often. Here is a compressed version of Sidoli's account:

"A new stage developed in Peter's play when he began to come to me twice a week. He identified with famous dictators and torturers, especially with Saddam Hussein. (It was the time of the Gulf War.) Several times I had to restrain his violent attacks on me. His attacks, both physical and verbal, were always accompanied by a great deal of farting. Whenever he felt like hating me – and in those days it was often – he would say that his foul smell was lethal gas with which he was poisoning me. Sometimes, however, he would be more ambivalent and less full of hate. Then he would warn me that the gas was on the way, and that I should wear a gas mask."

Later, Peter told her a story about pussycats, and the two engaged in a role-playing game. Peter the pussycat became demanding:

"I was ordered to stuff myself with great amounts of food that he would prepare for me. Then I had to pretend to vomit. At this stage in the therapy, he accompanied his game with miaowing, farting noises, and actual farts. His farts were especially loud when he was demonstrating how pussycat, after having eaten the whole world, would explode. Usually he had to rush to the toilet so as not to soil his pants."

This was a delicate stage in the therapy. As they entered the second year of working together, the relationship deteriorated. Peter was having a temporary regression – a regression of the kind that Jung always said was in service to the ego.

For Sidoli, these were the worst of times:

"The farts and farting noises, in spite of my constantly interpreting their meaning, increased in a manic crescendo. I began to feel impotent, unable to ever break through to him or alter the situation by verbal communication. My words seemed to bounce back at me, transformed into farts. The process was a perversion of transformation. He managed to turn my alpha-element interpretations back into beta-elements and then evacuate them back at me, preventing any processing of material to occur. I felt he had enveloped himself tightly into an invisible barrier of smells and farts that acted as protection against any external communication in the area of his greatest distress."

Sidoli recalled that Jung has faced a similar crisis point with one of his patients. Deciding to follow the path Jung had blazed, she launched a counter-attack, acting out her own games and making loud farting noises of her own.

This, Sidoli reports, produced the desired results:

"At first, Peter was very surprised and puzzled by my behaviour, not believing I could act in such a way. I took his surprise and puzzlement as a clue that he was beginning to hear me, and this encouraged me to keep on with this approach. His surprise was soon followed by annoyance, and then anger that I kept making such farting noises. He called me crazy and ordered me to stop. After a while, he gave me a long, steady look, and then burst out in whole-hearted laughter. At this I stopped. I told him that I had noticed that he used farting to stop communication, as if he wanted to make me believe he was crazy. I said that in the past he had managed to convince people that he was crazy, and had succeeded, in spite of being rejected because of it. After this, he began to relate to me like a 'real' child."

Eventually, Sidoli reports, "He reached a genuinely human form of expressing his dread and despair. Instead of enveloping me with farts, he was able to show me pain. He let out his love and his tears, and also his acute observations and his sense of humour."

The report concludes hopefully, as Peter has learned to relax his unbearable defence against unspeakable dread.

For her courage and tenacity in the face of a latency boy's defensive farting, and for assessing the experience with literary grace, Mara Sidoli won the 1998 Ig Nobel Prize in the field of Literature.

The winner could not attend the Ig Nobel Prize Ceremony, but expressed great pleasure at receiving a Prize. She said she had always accepted the most difficult psychiatric cases, and prided herself both on handling them well, and on writing them up with a measure of skill and style. In 1998, she became president of the National Association for the Advancement of Psychoanalysis.

The Apostrophe Protection Society

"Dear Sir or Madam," the letter begins, "Because there seems to be some doubt about the use of the apostrophe, we are taking the liberty of drawing your attention to an incorrect use."

–From the standard letter issued by the Apostrophe Protection Society

THE OFFICIAL CITATION

THE IG NOBEL LITERATURE PRIZE WAS AWARDED TO

John Richards of Boston, England, founder of The Apostrophe Protection Society, for his efforts to protect, promote, and defend the differences between plural and possessive.

The Apostrophe Protection Society can be reached at 23 Vauxhall Road, Boston, Lincs., PE21 0JB, United Kingdom.

After years of correcting his colleagues' errors, John Richards retired from the profession of newspaper subeditor. A lifetime of correcting erroneous apostrophe placement had both stoked his anger and prepared him for his true life's work.

After surviving a lengthy journalism career at the *Brighton Evening Argus*, the *Reading Evening Post*, the *Nottingham Evening Post*, the *West Sussex Gazette*, and the *West Sussex County Times*, Mr John Richards of Boston, Lincolnshire, had had his fill of seeing misused apostrophes.

Within weeks of his retirement, the anticipated happy days of relaxation were ruined by the sight of wayward apostrophes here, there, and all through the town. Mr Richards quickly reached what he called "boiling point." He would not let these outrages stand. Thus was born the Apostrophe Protection Society, an organization consisting, at first, of Mr Richards and his son Stephen. "I feel I am a lone voice – although my son agrees with me – but I shall carry on regardless," he said during this period.

IG NOBEL PRIZES

MIND YOUR APOSTROPHES!

Here is the Apostrophe Protection Society's concise guide to apostrophe usage.

The rules concerning the use of apostrophes in written English are very simple:

1. They are used to denote a missing letter or letters, for example:
I can't instead of **I cannot**
I don't instead of **I do not**
it's instead of **it is**

2. They are used to denote possession, for example:
• the **dog's** bone
• the **company's** logo
• **Jones's** bakery (but **Joneses'** bakery if owned by more than one Jones)

Note that we drop the apostrophe for the possessive form of it:
the bone is in **its** mouth
... however, if there are two or more dogs, companies or Joneses in our example, the apostrophe comes after the 's':
• the **dogs'** bones
• the **companies'** logos
• **Joneses'** bakeries

3. Apostrophes are never ever used to denote plurals! Common examples of such abuse (all seen in real life!) are:
• **Banana's for sale** which, of course, should read **Bananas for sale**
• **Menu's printed to order** which should read **Menus printed to order**
• **MOT's at this garage** which should read **MOTs at this garage**
• **1000's of bargains here!** which should read **1000s of bargains here!**
• **New CD's just in!** which should read **New CDs just in!**
• **Buy your Xmas tree's here!** which should read **Buy your Xmas trees here!**

NOTE: Special care must be taken over the use of **your** and **you're** as they sound the same, but are used quite differently:
• **your** is possessive, as in 'this is **your** pen'
• **you're** is short for **you are**, as in '**you're** coming over to my house'

Membership soon grew more than a hundredfold, and Mr Richards assumed the title of Chairman.

Chairman Richards told the *Daily Telegraph*: "It is something that has been irritating me for a long time. I walk around town and see so many misplaced or omitted apostrophes it beggars belief. The local fruiterer sells pounds of banana's, the public library, of all places, had a sign saying CD's – even Tesco was promising '1000's' of products at reduced prices. There were just so many of them, I thought something had to be done."

Chairman Richards strolls the streets of Boston, not aggressively searching for apostrophe problems, but prepared to deal with any he might encounter. The Apostrophe Protection Society has a form letter that covers virtually all contingencies. Its tone is polite, rather than hectoring: "We would like to emphasize that we do not intend any criticism, but are just reminding you of correct usage should you wish to put right the mistake."

One member of the Society "carries around sticky bits of tape with apostrophes on them, and sticks them on as required." But Chairman Richards uses no weapon other than the form letter and, on occasion, a sharply voiced rebuke.

The members of the Apostrophe Protection Society know they are fighting a war, not a battle, and that victory is unlikely during their lifetimes. Still, they soldier on, Chairman Richards leading the march.

For his way with the English language and for wanting his way with the English language, John Richards won the 2001 Ig Nobel Prize in the field of Literature.

The winner did not attend the Ig Nobel Prize Ceremony, as he does not like to travel on airplanes. Instead he sent a videotaped acceptance speech, in which he said:

"I must apologize for not being here in person, but I am surprised, even astonished and honored, to be making this acceptance speech here this evening. In fact, I can hardly believe it. It seems so unlikely, so improbable (which is an appropriate word to use in these circumstances), but I've felt deeply about the fate of the apostrophe for many years, and it's encouraged me to find that many thousands of people share my views. And it really beats me why so many people still maltreat this tiny, defenseless creature when, in the time it takes to listen to this speech, anyone can learn the basic rules of apostrophe usage by heart. In fact, I can hardly think of a better way of spending a minute, apart of course from listening to the acceptance speeches of our other guests. So now, thank you very much for the honor you've paid me this evening."

976 Co-Authors in Search of a Title

"I have no idea how many authors we have on that paper. I asked my assistant to count them, and she said she'd rather have a root canal."

–*New England Journal of Medicine* executive editor Marcia Angell

THE OFFICIAL CITATION

THE IG NOBEL LITERATURE PRIZE WAS AWARDED TO

E. Topol, R. Califf, F. Van de Werf, P. W. Armstrong, and their 972 co-authors, for publishing a medical research paper which has one hundred times as many authors as pages.

Their study was published as "An International Randomized Trial Comparing Four Thrombolytic Strategies for Acute Myocardial Infarction," *The New England Journal of Medicine*, vol. 329, no. 10, September 2, 1993, pp. 673–82.

In the scientific and medical worlds, having a published paper on the resumé can enhance one's prestige. It is not unusual for a paper to have two or more co-authors. It is not unusual to have five or even ten co-authors. To have 976 co-authors, though, is unusual.

A paper published in 1993 in the *New England Journal of Medicine* has approximately 976 co-authors. The word "approximately" is used here because observers who have tried counting the co-authors disagree as to the exact number, but the total is 976 plus or minus a few. The text of the medical report itself is just a few pages long. All told, the paper has one hundred times as many authors as pages.

The co-authors are in 15 different countries. It is doubtful that all of them have met each other. It is unclear whether even a single one has heard all the names read aloud. Nonetheless, they are co-authors.

For publishing this remarkable paper, the various and

sundry co-authors won the 1993 Ig Nobel Prize in the field of Literature.

The winners could not, or would not, attend the Ig Nobel Prize Ceremony, possibly because they could not agree on the wording of an acceptance speech. Had they come, the co-authors would have occupied more than two-thirds of the auditorium where the event was held. Dr Marcia Angell, the executive editor of the *New England Medical Journal of Medicine*, accepted the Prize on their behalf. Dr Angell said:

"On behalf of the *New England Journal of Medicine*, I accept, with dismay, this Prize. I have no idea how many authors we have on that paper. I asked my assistant to count them, and she said she'd rather have a root canal. I estimate there is one author for every two words in this article.

"This is all a part of our continuous author-enhancement campaign. The more papers you have, the more likely you are to be promoted and funded. If everyone can be an author on every paper, then everyone will be a tenured professor, and everyone will have a grant. So who can object to that?"

The (Approximately) 976 Co-Authors

Here is the complete list of co-authors of the medical research report that won the 1993 Ig Nobel Literature Prize. Some may be unaware that they are Ig Nobel Prize winners, so if you know any of these individuals, please inform them of their good fortune.

STEERING COMMITTEE: E. Topol (Study Chairman), United States; R. Califf (Clinical Director, Coordinating Center), United States; F. Van de Werf (Director, Intermediate Coordinating Center), Belgium; P.W. Armstrong, Canada; P. Aylward, Australia; G. Barbash, Israel; E. Bates, United States; A. Betriu,

Spain; J.P. Boissel, France; J. Chesebro, United States; J. Col, Belgium; D. de Bono, United Kingdom; J. Gore, United States; A. Guerci, United States; J. Hampton, United Kingdom; J. Hirsh, Canada; D. Holmes, United States; J. Horgan, Ireland; N. Kleiman, United States; V. Marder, United States; D. Morris, United States; M. Ohman, United States; M. Pfisterer, Switzerland; A. Ross, United States; W. Rutsch, Germany; Z. Sadowski, Poland; M. Simoons, Netherlands; A. Vahanian, France; W.D. Weaver, United States; H. White, New Zealand; and R. Wilcox, United Kingdom.

COORDINATING CENTER: DUKE UNIVERSITY MEDICAL CENTER, DURHAM, N.C.: CLINICIANS: R. Califf and G. Granger; STATISTICAL DIRECTOR: K. Lee; Statisticians: K. Pieper and L. Woodlief; Administrators: S. Karnash, J. Melton, and J. Snapp; COORDINATORS: L. Berdan, K. Davis, B. Hensley, C. Huffman, E. Kline-Rogers, J. Lee, I. Moffie, and D. Smith; PHARMACY: D. Christopher and M. Dorsey; PROGRAMMERS: C. Blackmon, B. Moss, and J. Shavendar; ON-CALL PHYSICIANS: R. Califf, C. Granger, B. Harrington, B. Hillegass, and M. Ohman.

EXECUTIVE CENTER: THE CLEVELAND CLINIC FOUNDATION, CLEVELAND: E. Topol, V. Stosik, D. Shyne, A. Thomas, D. Passmore, R. Wagner, D. Debowey, B. Keogh, and P. Brickenden.

INTERMEDIATE COORDINATING CENTER: UNIVERSITY OF LEUVEN, LEUVEN, BELGIUM: F. Van de Werf, I. Anastassiou, R. Brower, A. de Clerck, E. Lesaffre, A. Luyten, A. Meuris, P. Tenaerts, S. Van Dessel, and K. Verberckmoes.

AUSTRALIAN COORDINATING CENTRE: NATIONAL MEDICAL RESEARCH COUNCIL CLINICAL TRIALS

CENTRE, UNIVERSITY OF SYDNEY, SYDNEY, AUSTRALIA: J. Simes, E. Belles, S. Cho, J. Fabri, K. Farac, R. McCredie, and J. Sowden.

DATA AND SAFETY MONITORING BOARD: E. Braunwald (Chairman), M. Bertrand, M. Cheitlin, A. De Maria, D. De Mets, L. Fisher, P. Sleight, and L. Walters.

STROKE REVIEW COMMITTEE: N. Anderson, G. Barbash, J. Gore, P. Koudstaal, W. Longstreth, M. Simoons, M. Sloan, R. Tadmor, W.D. Weaver, and H. White.

UNITED STATES, NORTHEAST (CONNECTICUT, MASSACHUSETTS, MAINE, NEW HAMPSHIRE, VERMONT, NEW YORK, AND RHODE ISLAND): G. Macina, K. Salzsieder, C. Lambrew, R. Bishop, G. Gacioch, N. Jamal, J. Alexander, J. Layden, R. Grodman, J. DeSantis, H. Zarren, J. Cirbus, J. Morrison, D. Urbach, M. Capeless, E. Davison, G. MacDonald, B. Zola, G. Ryan, J. DiCola, J. Babb, W. Andrias, A. Binder, J. Robbins, P. Zwerner, M. Weinberg, J. Gore, C. Levick, A. Macina, R. Wallach, D. Miller, R. Kohn, A. Merliss, M. Falkoff, A. Sadaniantz, J. Greenberg, R. Parkes, W.H. Gaasch, S. Zeldis, L. Pinsky, M. Bakerman, B. Gaffney, M. Kaulbach, S. Labib, M. Therrien, A. Riba, J. Hanna, N. Brandon, S. Jacoby, H. Cabin, R. Dewey, D. Miller, J. Moses, A. Khan, V. LaDelia, R. Klare, H. Seidenstein, D. Losordo, M. Kukin, J. Strain, A. Rosenfeld, D. McCord, P. Bruno, P. Reiter, S. Blatt, A. Fass, A. Thomas, R. Shulman, B. Lindenberg, M. Bleiberg, J. Holbrook, M. Dharawat, J. Tumolo, S. Sheikh, G. Farrish, N. Niles, J. George, A. Sgalia, D. Parikh, E. Funk, C. Manning, E. Kosinski, R. Vince, H. Sanghvi, L. Sherman, J. Hsueh, F. Zugibe, L. Pisaniello, M. Sands, E. Pollak, Jr, E. Kehoe, M. Abdel-Azim, and B. Platt.

SOUTHEAST (NORTH CAROLINA, SOUTH CAROLINA, VIRGINIA, AND FLORIDA): J. McBride, P. Goodfield, M. Frey, P. Micale, E. Alsbrook, G. Miller, W. Maddox, R. Iwaoka, H. Morse, G. Pilcher, N. Trask, III, R. Jesse, M. Collins, J. Schrank, L. Howard, K. Sheikh, J. Puma, R. Califf, J. Barnes, B. Hearon, J. Dorchak, J. Kenerson, M. Johnson, J. Pasteriza, A. Magee, R. Schneider, C. Ashby, J. Nobel, M. Goldberg, J. Morris, S. Mester, W. Stuck, A. Rosenblat, G. Thomas, J. Smith, W. Ellison, W. Levy, M. Glover, D. Eich, P. Popper, K. Gibbs, R. Seagle, G. Lane, K. Popio, A. Blaker, A. Tse, D. McMillan, R. Vicari, A. Whitaker, D. Mokotoff, S. Roark, D. Ike, A. Ghahramani, C. Davenport, J. Hoekstra, D. Givens, R. Dunkelberg, R. Schneider, M. Clark, F. Lenz, M. Whisenant, M. Lopez, S. Schnider, J. Strickland, R. Palaniyandi, R. Stack, A. Bartel, T. Long, E. Hawkins, R. Everhart, R. Goulah, R. Lewis, R. Thigpen, S. West, J. Anderson, M. Hajisheik, and D. Privette.

GREAT LAKES (INDIANA, KENTUCKY, MICHIGAN, AND OHIO): R. Josephson, R. Schumacher, K. Mohan, G. Litman, J. Formolo, D. Besley, A. Klaus, L. Calli, Jr., W. Duvernoy, J. Heinsimer, J. Schaeffer, R. Miller, R. Stomel, E. Papasifakis, M. Zande, J. Jacobs, J. Kazmierski, K. Holland, F. Griff, W. Whitaker, S. Weinberg, J. VanGilder, J. Rogers, D. Dageford, P. Bacidore, M. Rubin, R. Reynolds, A. Razavi, J. Hodgson, R. Millsaps, F. Wefald, T. Fraker, Jr., R. Vanderlaan, K. Scully, B. Morrice, J. Forchetti, R. Kurtz, W. Meengs, A. Weizenberg, M. Tejura, E. Bates, P. Fleisher, B. Perry, M. Kreindel, D. Kereiakes, T. Vrobel, M. James, E. Basse, P. Andres, B. Lew, S. Zampani, M. Gheorghiade, C. Milford, W. Wilson, S. Bhatia, T. Doyle, S. Traughber, W. Polinski, S. Brownstein, E. Topol, M. Meyer, T. Heft, K. Kuppler, B. Schilt, V. Mistry, and D. Booth.

MID-ATLANTIC (DISTRICT OF COLUMBIA, DELAWARE, MARYLAND, NEW JERSEY, PENNSYLVANIA, AND WEST VIRGINIA): R. Bahr, A. Doorey, T. Krisanda, J. Smith, R. Biern, J. Gregory, N. Strahan, A. Bramowitz, R. Gordon, J. Ibarra, A. Ross, S. Worley, W. Berkowitz, R. Fields, M. Effron, K. Lindgren, E. Roseff, M. Avington, S. Sharma, J. Banas, W. Beckwith, V. Krishnaswami, T. Boyek, H. Dale, J. Zimmerman, J. Burks, L. Gehl, A. Meshkov, R. Rubinstein, G. Groman, J. Ellis, IV, A. Popkave, D. Ferri, M. Santer, Jr., L. Konecke, K. Singal, J. Wertheimer, H. Selinger, M. Borsch, H. Starr, T. Parris, M. Pecora, J. Patankar, W. Noble, G. Grossman, B. Clemson, D. Rosing, L. Denlinger, L. Adler, H. Goldschmidt, J. O'Toole, D. McCormick, J. Granato, C. Naganna, E. Gerber, T. Little, R. Angeli, W. Markson, O. Randall, M. Kesselbrenner, K. Olsen, W. Esper, and K. Hawthorne.

SOUTHWEST (ARIZONA, COLORADO, IDAHO, KANSAS, MONTANA, NEW MEXICO, TEXAS, UTAH, AND WYOMING): M. Padnick, H. White, Jr., M. Stern, T. Lombardo, J. Svinarich, P. Browne, J. Saini, N. Laufer, S. Ung, D. Rigby, J. Perry, A. Mattern, N. Shadoff, V. Aquino, A. Newton, L. Lancaster, D. Gonzalez, G. Symkoviak, W. Falcone, N. Israel, R. Scott, G. Hui, J. Boerner, K. Nademanee, J. Sbarbaro, M. Kraus, H. Lee, D. Sellers, B. Owens, S. Harris, D. Brown, M. Solovay, A. Damien, S. Woolbert, B. Call, M. McGuire, T. Glatter, R. Davis, E. Terry, C. Castle, R. Oliveros, J. Laser, C. Haws, R. Park, F. Cecena, C. Dahl, S. Gollub, R. Heuser, G. Peese, M. Sanz, C. Brooks, C. Schechter, J. Gladden, R. Bond, M. Crawford, R. Loge, J. Moreland, L. Faitelson, W. Lewis, R. Dattilo, M. Carbajal, R. Tabbaa, G. Rodgers, J. Morgan, M. Traylor, C. Unrein, R. Crossno, and C. Wilkins.

MIDWEST (ILLINOIS, WISCONSIN, MISSOURI, SOUTH DAKOTA, NORTH DAKOTA, NEBRASKA, IOWA, AND MINNESOTA): G. Hanovich, W. Hession, B. Abramowitz, J. Thompson, S. Kopecky, L. Cook, J. Drozda, L. Swenson, P. Schmidt, A. Mooss, B. Anderson, D. Goldsteen, F. Ferrigni, A. Edin, C. Santolin, J. Alexander, K. Fullin, J. McCriskin, G. Taylor, D. Shuster, L. Solberg, R. Menning, L. Abrahams, J. Epplin, S. Benton, B. Handler, N. Streitmatter, M. Saddin, W. Lam, I. Silverman, R. Dinter, W. Frank, D. Zwicke, D. Pfefferkorn, T. Matzura, D. Meyers, S. Bloom, C. Jones, P. Quandt, M. Wheeler, C. Monroe, D. Jenny, H. Coleman, R. Holm, L. Shelhamer, Jr, G. Grix, P. Leutmer, R. Harner, C. Koeppl, R. Yawn, P. Anantachai, K. Jaeger, B. Patel, M. Cinquegrani, T. Dynes, C. Campanella, D. Larson, S. Gill, C. Thompson, K. Kavanaugh, N. Harb, D. Dixon, J. Carr, J. Shanes, V. Miscia, A. Hsieh, and R. Pensinger.

WEST (ALASKA, CALIFORNIA, HAWAII, NEVADA, OREGON, AND WASHINGTON): P. Lightfoot, R. Swenson, P. Sarkaria, R. Acheatel, J. Rudoff, R. Anschuetz, E. Lapin, R. Spiegel, P. Lai, B. Strunk, W. Rowe, R. Finegan, B. Gross, J. Chappell, T. Berndt, B. Titus, R. Oikawa, R. Ashmore, D. Bayne, G. Wesley, E. Quinn, K. Jutzy, G. Fehrenbacher, P. Kotha, P. Phillips, K. Ryman, J. Holmes, H. Kwee, D. Cislowski, R. Bream, T. Elder, III, H. Olson, R. Trenouth, C. Wolfe, S. Raskin, J. Comazzi, K. Stokke, M. Nallasivan, D. Hogle, B. Kennelly, E. Wroblewski, J. Altamirano, E. Chesne, A. Choe, and A. Brodersen.

MIDDLE SOUTH (ALABAMA, GEORGIA, LOUISIANA, OKLAHOMA, TENNESSEE, ARKANSAS, MISSISSIPPI, AND FLORIDA): S. Sherman, E. Pickering, J. Kalbfleisch, C. Williams, J. Dedonis, M. Silverman, M. Geer, K. Wright, D. Williams,

W. Guest, R. Sinyard, Jr., Z. Baber, S. Howell, III, R. Ingram, D. Morris, W. Beeson, R. Schlant, V. McLaughlin, H. Hanley, G. Olson, P. Gainey, D. Shonkoff, Y. Ong, G. Phillips, F. Kushner, C. White, J. Hoopes, P. Breaux, J. Lam, M. Honan, R. Hill, M. Certain, H. Ba'abaki, T. Atha, H. Butler, L. Battey, J. Scott, G. Cash, P. Mullen, R. Wrenn, A. DeLeon, U. Thadani, L. Price, E. Magiros, and P. Subramaniam.

ISRAEL: H. Hamerman, D. David, S. Sklerovsky, G. Barbash, B. Peled, S. Laniado, N. Rogin, S. Schlezinger, I. Zehavi, A. Caspi, E. Barash, Y. Kishon, A. Keren, A. Palant, E. Avineder, T. Weis, L. Reisin, D. Zivony, L. Rudnik, B. Luis, A. Marmur, M. Gotesman, and E. Gelvan.

CANADA: S. Roth, D. Roth, M. Traboulsi, M. Henderson, K. Finnie, J. Burton, R. Trifts, J. McDowell, P. Klinke, R. Lesoway, M. Senaratne, B. Lubelsky, E. Goode, M. Cheung, P. Bogaty, B. Burke, C. Morgan, M. Turek, A. Hess, C. Lefkowitz, J. Charles, P. Armstrong, A. Fung, G. Kuruvilla, D. Langleben, B. Hrycyshyn, C. Kells, R. Delarochelliere, V. Sluzar, K. Kwok, M. Goddard, J. Fulop, J. Brophy, A. Zawadowski, B. Sahay, F. Ervin, C. Thompson, A. Abdulla, K. Boroomand, C. McMillan, P. Carter, P. Laramee, R. Hathaway, M. O'Reilly, S. Vizel, D. Hilton, G. Jablonsky, P. Bolduc, L. Simard, N. Ranganathan, D. Gould, L. Bate, D. Cameron, B. Mackenzie, P. Greenwood, D. Gossard, J. Blakely, J. Morch, R. Mildenberger, N. Racine, and H. Baillie.

NETHERLANDS: A.E.R. Arnold, J.G. Engbers, B.J.L. DeRode, G.P. Molhoek, P.M. Van Kalmthout, L. Cozijnsen, C.L. Van Engelen, J.H.M. Deppenbroek, S.K. Oei, J.B.L. ten Kate, M.J. de Leeuw, G.J. Laarman, J.V.C. Stevens, D. Haan, L. van Bogerijen, W.C.G. Smits, P.W. Westerhof, P.W.J. Stolwijk,

H.A.M. Spierenburg, E.J. Muller, B. Cernohorsky, J.J.J. Bucx, H.J.A.M. Penn, H. Fintelman, C. Van Rees, M.L. Simoons, J. Kerker, E.G. Faber, R. Bergshoeff, H.W.O. Roeters Van Lennep, W. Muys v/d Moer, L. Relik-van Wely, F. van Bemmel, R.J. Bos, A. Zwiers, C.M. Leenders, P. Zijnen, D.G. de Waal-Ultee, H. De Rebel-De Vries, S.A.G.J. Witteveen, and P. de Weerd.

AUSTRALIA: P. Aylward, B. Hockings, M. Brown, D. Cross, G. Lane, G. Aroney, D. Hunt, B. Singh, A. Tonkin, P. Thompson, G. Nelson, R. Newman, J. Federman, T. Campbell, J. Healey, D. Ramsey, W. Ryan, J. Counsell, D. Coles, A. Thomson, S. Woodhouse, G. Simmons, P. Harris, P. Caspari, A. Limaye, T. Donald, S. Coverdale, G. Smith, R. Walker, R. Harper, C. Gnanharan, P. Carroll, J. Woods, C. Hadfield, P. French, A. Groessler, B. Morphett, G. Phelps, B. Quinn, K. Gunawardane, P. Kertes, C. Medley, A. Soward, T. Htut, A. Appelbe, J. Johns, I. Beinart, R. Hynes, M. Knapp, P. Curteis, D. Owensby, P. Davidson, W. Renton, P. Windsor, L. Bolitho, B. Forge, R. Ziffer, R. McLeay, R. Cranswick, and L. Mollison.

BELGIUM: H. De Geest, F. Van de Werf, G. Verstreken, J. Col, R. Beeuwsaert, J. Boland, A. Vanrossum, H. Lesseliers, R. Popeye, Ph. Dejaegher, B. Pirenne, E. Van der Stichele, J. Chaudron, M. Castadot, L. Dermauw, G. Vanquickenborne, W. Van Meghem, H. Robijns, M. Vankuyk, C. Emmerechts, D. Dendooven, H. Van Brabandt, E. Installe, S. Dierickx, C. Haseldonckx, H. Lignian, J. Beys, P. Noyens, A. Van Dorpe, Ph. Henry, Ph. Van Iseghem, F. Gielen, D. Lanoy, P. DeCeuninck, J. Schurmans, L. Geutjens, M. Carlier, P. Surmont, Ch. Henuzet, P. Van Robays, R. Stroobandt, P. Peerenboom, C. Mortier, X. Dalle, K. Mitri, U. Van Walleghem, J. Bonte, D. Koentges, A. De Paepe, L. DeWolf, Th. Sottiaux, J. Van Besien, P. Van den Heuvel, H. Ulrichts, Y. Deheneffe, H. Jacobs, J. Croonenberghs,

L. Pirot, J. Carpentier, R. Schreuer, L. Vermeersch, D. Stroobants, D. Missotten, E. Marchand, S. De Schepper, B. Carlier, Ch. Doyen, A. Palmer, M. Jottrand, C. Gillebert, M. Bayart, A. Van Wylick, J. Leonard, and E. Benit.

GERMANY: W. Rutsch, H. Topp, H. Simon, H. Ditter, P. Wylicil, H. Meyer-Hoffmann, H.P. Nast, R. Engberding, K. Caesar, U. Schmitz, W. Jansen, H.R. Ewers, H.U. Kreft, D. Kaut, P. Schweizer, J. Cyran, U. Peters, E. Horstmann, R. Koch, R. Scheemann, J. Bolte, W. Berges, K.P. Schueren, M.H. Hust, H.U. Koch, W. Overbuschmann, B. Henkel, S. Troost, R. Jacksch, W. Burkhardt, H. Loellgen, J. Schimanski, H. Callsen, P.W. Kummerhoff, H. Hochrein, G.M. Mueller, H. Schulz, V. Hossmann, F. Voehringer, D. Boettcher, P. Glogner, K.H. Hohmann, H.J. Von Mengden, W. Krengel, B. Maisch, P. Spiller, M. Adamczak, R. Wacker, W. Urbaszek, H.D. Bundschu, W. Ernst, R. Eisenreich, M. Konz, C. Dienst, J.G. Schmailzl, A. Gartemann, W. Sill, C. Piper, J. Schiffner, N. Meyer-Guenther, D. Siebenlist, E. Chorianapoulos, R. Schroeder, P. Oehl, W. Lengfelder, J. Djonlagic, H.W. Hopp, W. Weser, P. Kahl, P.H. Althoff, R. Hopf, R. Oberheiden, H.V. Lilienfeld-Toal, G. Schulte-Herbrueggen, and P. Doenecke.

FRANCE: J. Valty, A. Py, J. Acar, A. Vahanian, G. Grollier, D. Barreau, K. Khalife, J.C. Quiret, X. Tran Thanh, J.P. Bourdarias, P. Besse, M. Hiltgen, P. Bernadet, J. Boschat, C. Gully, J.M. Mossard, B. Charbonnier, F. Funck, M. Bedossa, R. Grolleau-Raoux, J. Cassagnes, J.C. Daubert, Ph. Beaufils, J.M. Juliard, G. Bessede, B. Vitoux, C. Thery, G. Hanania, C. Mycinski, E. Brochet, C. Cassat, C. Socolovsky, R. Mossaz, J.L. Fincker, M. Lang, J.L. Guermonprez, J.M. Demarcq, A. Page, C. Guerot, R. Barraine, Ph. Morand, A. Bajolet, J. Vedel, P. Dambrine, H. Lardoux, B. Veyre, A. Vacheron, F. Latour, J.P. Normand, J.Y. Thisse, J. Machecourt, J.P. Bassand,

B. Carette, C. Toussaint, J.P. Cebron, M.F. Bragard, Ph. Geslin, O. Leroy, G. Allard-Latour, F. Fockenier, J. Gauthier, M. Escande, and M. Viallet.

UNITED KINGDOM: R.G. Wilcox, R.D. Thomas, R.M. Boyle, R.H. Smith, E.T.L. Davies, J. Kooner, G. Terry, B. Gould, M.O. Coupe, J.E.F. Pohl, E.W. Barnes, H. Simpson, A. Davis, J.A. Bell, I.N. Findlay, P. Wilkinson, G.C. Sutton, T.S. Callaghan, E.J. Wakely, D. Waller, G. Tildesley, and R.L. Blandford.

NEW ZEALAND: P. Leslie, H. Ikram, S. Foy, S. Mann, A. Mylius, S. Anandaraja, M. Singh, D. Friedlander, B. Bruns, L. Nairn, M. Abernethy, R. Rankin, D. Durham, J. Doran, M. Audeau, H. White, S. Reuben, G. Lewis, H. Hart, and G. Wilkins.

SPAIN: G. Froufe, X. Bosch, F.F. Aviles, C.M. Luengo, J. Corrons, L.L. Bescos, A. Loma, R. Masia, J. Figueras, V. Valle, L. Saenz, A. Betriu, E. Alegria, and J. Eizaguirre.

POLAND: T. Kraska, J. Kuch, A. Dyduszynski, J. Stepinska, K. Wrabec, E. Czestochowska, Z. Sadowski, K. Zawilska, Z. Kornacewicz-Jach, E. Nartowicz, W. Piwowarska, G. Swiatecka, J. Wodniecki, T. Petelenz, and A. Kalicinski.

SWITZERLAND: P. Urban, M. Pfisterer, O. Bertel, H.R. Jenzer, W. Angehrn, and H.R. Baur.

IRELAND: K.M. Daly, J. Horgan, M. Walsh, J. Taaffe, D. Murray, D. Sugrue, P. Sullivan, B.C. Muldoon, D. McCoy, B. Maurer, G. Fitzgerald, T. Pierce, and K. Balnave.

LUXEMBOURG: R. Erpelding.

948 Titles in Search of an Author

"One may ask why Yuri forced publication so hard, what was the motivation that did not permit him to make his mission easier?"

–from the obituary of Yuri Timofeevich Struchkov (1926–1995) written by Alajos Kálmán and published in the journal *Acta Crystallographica*

THE OFFICIAL CITATION

THE IG NOBEL LITERATURE PRIZE WAS AWARDED TO

Yuri Struchkov, unstoppable author from the Institute of Organoelemental Compounds in Moscow, for the 948 scientific papers he published between the years 1981 and 1990, averaging more than one every 3.9 days.

For many academic scientists, having a list of published papers is the single most important factor in determining prestige, pay, promotions, and job offers. Some scientists are more prolific than others. But one man – Yuri T. Struchkov – established a record of almost superhuman accomplishment. During a ten-year span he published more – far more – research papers than any other scientist on earth.

Yuri T. Struchkov was director of the Institute of Organoelemental Compounds of the Academy of Sciences, in Moscow. He was one of the world's great crystallographers. A crystallographer uses X-ray machines to take pictures of crystals. In the 20th century, this was one of the most important and powerful techniques by which chemists learned about the structures of complex molecules. Many a chemist has won honor and glory – and in some cases Nobel Prizes – by using crystallography to explore nature's chemical secrets.

Still, for all his renown in the crystallographic community, Yuri Struchkov was virtually unknown to the world in general until 1992. That year David Pendlebury, of the Institute for Scientific Information, in Philadephia, used the Institute's massive science citation database to see which scientist had published the largest number of papers during the decade 1981–1990.

The winner, hands down, was the little-known Dr Struchkov. 948 research reports listed Yuri T. Struchkov as either author or co-author. On average, a new Struchkov paper was published every 3.9 days during the entire ten-year span.

Virtually all of these papers are in the field of crystallography, and almost all list Struchkov as one of several co-authors, rather than as sole author.

During 1985, for example, he was co-author with Pyshnograeva, Batsanov, Ginzburg, and Setkina of the intriguing treatise "Basicity of Metal-Carbonyl Complexes .19. CO Substitution in Azacymantrene and Reactions of (ETA-C_4H_4N)$MN(CO)_2PPH_3$ with Electrophiles – X-Ray Crystal-Structure of [(PPH_3)$(CO)_2$MN(ETA-5-C_4 H_4N)$]_2$ $PDCL_2$."

That same year he was co-author with Cherepinskiimalov Gurarii Li, Mukmenev, and Arbuzov, of the much-commented-upon disquisition "Crystal and Molecular-Structures of Diethylammonium Salt of 2-Hydroxy-4,5-Dibromophenyl Phosphoric-Acid, C10H16BR2N O5P."

He also co-authored, with Egorov, Kolesnikov, Antipin, Sereda, and Nefedov, "Structure of Delta-1.7-2,2,6,6-Tetramethyl-4-Thia-8,8-Dimethyl-8-Germabi-Cyclo[5.1.0] Octene - The 1st Example of a Germacyclopropene."

Those, of course, are just three of the 90 or so papers he co-authored that year, and 1985 was just one year out of ten. It should be noted that Struchkov's golden run began well before 1981, and continued well after 1990.

For his prodigious contributions to world literature, Yuri

T. Struchkov won the 1992 Ig Nobel Prize in the field of Literature.

The winner could not, or would not, attend the Ig Nobel Prize Ceremony.

Yuri Struchkov continued to publish scientific papers at a torrid pace.

Struchkov died in 1995. An obituary published in the journal *Acta Crystallographica* said that altogether he had published more than 2,000 scientific papers. The obituary's author marveled at Struchkov's output, and hazarded a guess as to what motivated him:

"One may ask why Yuri forced publication so hard, what was the motivation that did not permit him to make his mission easier? His most convincing argument for not joining the Communist Party was his deep preoccupation with research which left no time for anything else. He felt his only choice was to work hard, harder today than yesterday and much harder tomorrow. He was awarded the A.N. Nesmeyanov Gold Medal only in 1988 and elected belatedly

IG NOBEL PRIZES

DEATH IS NO IMPEDIMENT

Yuri Struchkov.

Although Yuri T. Struchkov died in 1995, he continued to publish scientific research papers. The paper shown overleaf, for example, was submitted to the journal *Organometallics* in 1998 and published in 1999, with Struchkov listed as the seventh of nine co-authors. It is just one of many Struchkov co-authored papers published subsequent to his passing. In terms of output, the late Dr Struchkov was still going strong even after the turn of the century.

Organometallics **1999**, *18*, 726–735

Synthesis of Mixed-Metal (Ru–Rh) Bimetallacarboranes via *exo-nido-* and *closo-*Ruthenacarboranes. Molecular Structures of (η^4-C_8H_{12})Rh(μ-H)Ru(PPh_3)$_2$(η^5-$C_2B_9H_{11}$) and (CO)(PPh_3)Rh(μ-H)Ru(PPh_3)$_2$(η^5-$C_2B_9H_{11}$) and Their Anionic *closo-*Ruthenacarborane Precursors

Igor T. Chizhevsky,* Irina A. Lobanova, Pavel V. Petrovskii, Vladimir I. Bregadze, Fyodor M. Dolgushin, Alexandr I. Yanovsky, Yuri T. Struchkov,† Anatolii L. Chistyakov, and Ivan V. Stankevich

A. N. Nesmeyanov Institute of Organoelement Compounds, 28 Vavilov Street, 117813 Moscow, Russian Federation

Carolyn B. Knobler and M. Frederick Hawthorne*

Department of Chemistry and Biochemistry, University of California at Los Angeles, Los Angeles, California 90095-1569

Received July 31, 1998

The reaction of *exo-nido-*5,6,10-[Cl(PPh_3)$_2$Ru]-5,6,10-(μ-H)$_3$-10-H-7,8-$C_2B_9H_8$) (**1**) with [(η^4-diene)RhCl]$_2$ in EtOH or with [(CO)$_2$RhCl]$_2$ in MeOH in the presence of KOH produced novel mixed-metal bimetallacarboranes (η^4-diene)Rh(μ-H)Ru(PPh_3)$_2$(η^5-$C_2B_9H_{11}$) (**4**, diene = COD; **5**, diene = NBD) or (CO)(PPh_3)Rh(μ-H)Ru(PPh_3)$_2$(η^5-$C_2B_9H_{11}$) (**10**) along with a small amount of the mononuclear complexes *closo-*(CO)$_2$(PPh_3)Ru(η^5-$C_2B_9H_{11}$) and *closo-*(CO)(PPh_3)$_2$Ru-

This research paper appeared much too late – it was published long after Struchkov's death, in fact – to be among the 948 that earned Struchkov his prize.

as a Corresponding Member of the Academy of Sciences in 1990."

That's one possibility. Another was mentioned by Western scientists who were familiar with the situation in the USSR. The equipment to do proper crystallography was scarce in the Soviet Union. It was rumored that Soviet scientists who needed to do crystallography were welcome to use the equipment at the Institute of Organoelemental Compounds of the Academy of Sciences, in return for adding a certain person to the list of co-authors when it came time to write up their results.

Chariots of the Gods

"It took courage to write this book, and it will take courage to read it."
–the first words in the book *Chariots of the Gods*?

THE OFFICIAL CITATION

THE IG NOBEL LITERATURE PRIZE WAS AWARDED TO

Erich Von Däniken, visionary raconteur and author of *Chariots of the Gods*?, for explaining how human civilization was influenced by ancient astronauts from outer space.

The book was published in 1968 by Putnam Press and Bantam Books.

In 1968, Erich Von Däniken arrived as if on a chariot of gold, bringing inspiration and riches to the book industry. His first book touched off a frenzy, and gave birth to the business of studying ancient astronauts from outer space.

As managing director of a Swiss hotel, Erich von Däniken often dealt with guests who arrived mysteriously in the dark and vanished leaving only cryptic, but alluring, traces. And thus, von Däniken was well qualified to write the book *Chariots of the Gods?* The book was published in 1968 in the United States and Germany, and Von Däniken's life changed drastically. He never again worked in a hotel.

Chariots of the Gods? shows that history is not the collection-of-dull-facts-and-pat-explanations subject taught in schools. One sentence above all rings out from Von Däniken's text: "None of these explanations stands up to a critical assessment."

Chariots of the Gods? showed that history is all critical assessments. And even more, it's all about mysteries, such as:

• The mystery of the ancient airport landing strips on the plain of Nazca in Peru.

• The mystery of the detailed drawing, found in the ruins of the Mayan city of Copán, of an astronaut working the controls of a rocket, with flames and gases emerging from the propulsion unit below him; and the mystery of the Assyrian cylinder from the 1st century BC that shows a symbolic representation of the construction of the atom, and also shows an astronaut piloting a fiery chariot.
• The mystery of the ancient drawing in Val Camonica in Northern Italy which shows the extraordinary obsession primitive man had with figures in spacesuits and unusual headgear.
• The mystery of the ancient sonar map, found in the Topkapi Palace in Istanbul, showing the conformation of the land mass that lies beneath the miles of ice and snow of Antarctica, and which is invisible to the eyes of man.
• The mystery of the sophisticated electrical batteries, now dead, found in Bronze Age cities, and the mystery of the electrical wiring of the Ark of the Covenant described in the Bible.

Those are just a few of the mysteries.

The book has mysteriously been translated into 28 languages, and has mysteriously sold many millions of copies. And there are subtle mysteries concerning the book itself. The book's length, for example, is a mystery that has never been fully explained. The first edition of *Chariots of the Gods?* was 189 pages long. A subsequent edition had 163 pages; yet another edition had 169. And observers report having seen editions with still other lengths.

The success of the book led to a television special called *In Search of Ancient Astronauts*, which attracted a large audience, and led to the making of many other TV documentaries. It also led to a number of wildly speculative videos, movies, and books by people who did not have Erich

Erich Von Däniken's Prize-winning book.

Von Däniken's scholarly care about getting things right, and who may have been more interested in making money than in plumbing the very depths of the unknown.

Von Däniken himself was not content to rest on his laurels and royalties. He kept at his investigations, which he documented over the next 20 years in a stream of scholarly publications: *Gods From Outer Space*; *The Gold of The Gods*; *In Search of Ancient Gods*; *Miracles of The Gods*; *Von Däniken's Proof*; *Im Kreuzverhör*; *Signs of the Gods?*; *Pathways to the Gods*; *The Gods and Their Grand Design*; *Ich Liebe Die Ganze Welt*; *Der Tag An Dem Die Götter Kamen*; *Habe Ich Mich Geirrt?*; *Wir Alle Sind Kinder Der Götter*; *Die Spuren Der Außerirdischen*; *Die Steinzeit War Ganz Anders*; and *Die Rätsel im Alten Europa*.

Von Däniken's curiosity about ancient astronauts from outer space was and is insatiable. His reputation as a scholar is firmly established. He always respects the boundaries of common sense, and keeps them at a distance.

"I know that astronauts visited the earth in ancient times," he told the respected journal *The National Enquirer* in 1974, because "I was there when the astronauts arrived. And I know they'll be back."

For challenging his readers again and again to buy his books, Erich Von Däniken won the 1991 Ig Nobel Prize in the field of Literature.

The winner could not, or would not, attend the Ig Nobel Prize Ceremony.

Winning the Prize did not dim his enthusiasm or slow Erich Von Däniken's drive for profitable exploration. After 1991, he continued to do research in his chosen field, and published a corpus of further works that include: *Der Götterschock*; *Raumfahrt im Altertum*; *Das Erbe von Kukulkan*; *Auf den Spuren der All-Mächtigen*; *The Eyes of the Sphinx; The Return of the Gods – Evidence of Extraterrestrial Visitations*; *The Arrival of the Gods – Revealing the Alien Landing Sites of Nazca*; and *Odyssey of the Gods – An Alien History of Ancient Greece.*

In 2002, Von Däniken and a consortium of investors (including Feldschlösschen, Valora, Sennheiser, Sony, Hotela, Coca-Cola, Bleuel Electronic, and Hewlett Packard) launched the Mysteries of the World Theme Park in Interlaken, Switzerland, near the author's home. Two of the early advertising slogans were: "Close encounter in the home land of Erich Von Däniken!" and "Interlaken will become the Mecca of great puzzles of the world."

The Father of Junk E-Mail

"Sanford Wallace is laughing today. The man is in his element, phoning reporters from his cellular phone about that cease-and-desist letter sent to him by Hormel Foods Corporation ordering him to stop using a certain word to promote his business, which is bombarding the Net with unsolicited junk e-mail.

"You know the word: 'spam.' Wallace isn't about to stop using it. It's even less likely he'll stop using it now, since his use of the word 'spam' has put him directly in the spotlight, where he is happiest. In fact, he couldn't have paid enough for this publicity. If he weren't in a legal dispute with Hormel, he'd probably send them a thank-you e-mail note – unsolicited of course."

–from a report by *CNET News*, July 2, 1997

THE OFFICIAL CITATION

THE IG NOBEL COMMUNICATIONS PRIZE WAS AWARDED TO

Sanford Wallace, president of Cyber Promotions of Philadelphia – neither rain nor sleet nor dark of night have stayed this self-appointed courier from delivering electronic junk mail to all the world.

Sanford Wallace was called the "King of Spam," "The Most Reviled Man on the Internet," and several thousand other names, all expressing a similar type of admiration. Wallace earned the appellations and the revulsion. Through his efforts, junk e-mail messages became more numerous, in many parts of the earth, than rats, mice or cockroaches.

Sanford Wallace started a business to help people use the then-fairly-new medium of electronic mail. Realizing that it was beyond his ability to help everyone, Wallace concentrated on helping people who wanted to send junky advertisements and who were willing to pay an amoral stranger to do it for them.

In the pre-Wallace days on the Internet, there was a widely honored code of good behavior about sending e-mail. The rule of thumb was: don't send unsolicited advertising to strangers.

Because it was so simple and inexpensive to send a message – or a dozen messages, or even several thousand – everyone recognized the potential for annoyance and abuse. Some newcomers to the Internet, unfamiliar with the etiquette, would send out unsolicited ads. Generally they would receive in return an instant flood of complaints, polite and otherwise, and typically they would then temper their behavior.

Sanford Wallace understood the potential for annoyance and abuse. He recognized that in the large numbers of people using e-mail there lay opportunity. If he were to e-mail an ad to 10,000, strangers, one or two might purchase whatever was being advertised. Or maybe one or two people out of a 100,000 would make a purchase. The fact that 9,999 or 99,999 people might be annoyed was not relevant. And so Sanford Wallace solicited some advertisers and began sending ads on their behalf to thousands upon thousands of strangers.

When the complaints came in, which always happened within minutes of his sending out a new wave of ads, he ignored them. In a surprisingly short amount of time, hundreds of thousands of opportunistic companies were paying Cyber Promotions to send out electronic junk mail for them. And every day millions of people were finding unsolicited junk mail, often in large quantities, flooding into their e-mail accounts.

"If you want to use junk e-mail," Wallace told potential customers, "the bottom line is that it works. It gets results."

Wallace claimed to be sending more than 15 to 20 million pieces of spam a day, and no one doubted his claim. All across the Internet, people came to hate and dread the junk e-mail that poured in on them, dozens or even hundreds of messages day after day, without cease. They loathed spending the time it took to separate their real e-mail messages from the trash that came in with it – the Get Rich Quick schemes, the penis enlargers, the pornographic videotapes, the cut-rate household implements, the fake diplomas from real colleges and real diplomas from fake universities, and so much more, ever so much more, much too much more.

In no time at all, and for reasons that are not all clear, the word "spam" somehow became the universal term for junk e-mail. The act of sending spam was called spamming. Sanford Wallace announced that he loved being called "The King of Spam" and would further enjoy it if people were to call him "Spamford Wallace." The Hormel Company, manufacturer of the curious foodstuff called Spam (the consumers of which were honored with an Ig Nobel Prize in 1992, "for 54 years of undiscriminating digestion"), became enraged at the hijacking of their product's good, greasy name – and especially at Sanford Wallace's boastful trumpeting of it – but discovered that legally they could not put a can on this new use of spam.

Sanford Wallace's business was lucrative and increasingly famous, but it did have problems.

To be able to send e-mail, Cyber Promotions had to have e-mail accounts. Although several thousand companies were in the business of providing such accounts, few were willing to have Cyber Promotions as a customer. Whenever he would begin using a new Internet service provider, it took just hours for outraged spam recipients to track down the company and inundate it with complaints, threats, and lawsuits.

This being the Internet, the technical adepts had their own ways of expressing displeasure. Hackers hacked Cyber Promotion's web site, crippling it time and again. These were anonymously altruistic, if not quite fully law-abiding, programmers who believed it would be a very public good deed to shut down a very public nuisance.

For a surprisingly long time, Sanford Wallace was unfazed by the difficulties. They were simply the cost of doing business, and they were making him a big, albeit detested, name. No matter how many Internet service providers booted Cyber Promotions off their systems, Wallace always managed to find new ones who would put up with him for a day or two or three, long enough for him to spew out several million more e-mails and find yet another Internet service provider.

For his successful and voluminous efforts to spam the globe, Sanford Wallace won the 1997 Ig Nobel Prize in the field of Communications.

The winner did not attend the Ig Nobel Prize Ceremony. The Ig Nobel Board of Governors did not invite him, fearing for the man's safety.

Over the next few months, Cyber Promotions kept up the good fight for the right to send annoying, unsolicited e-mail to anyone, anywhere, any time. Inspired by Wallace's chutzpah, publicity, and client base, competitors multiplied, very much in the manner of cockroaches, mice, and rats.

The incessant lawsuits and the competition together proved too much for the Spam King. In 1998, he abandoned Cyber Promotions. At the same time, he professed to have a change of heart. *Wired News* reported on the spectacle:

"'I will never spam again. Period. I will never be affiliated with spam again. Period.' With those words, Sanford

Wallace, the 29-year-old bad boy of the Internet, officially renounced his title as the King of Spam and promised to be good from now on. Not only will he behave, he will announce his support for the Smith Bill – aka HR 1748, the Netizens Protection Act – federal legislation now before Congress which would outlaw spam … As recently as six months ago, Wallace was saying he saw nothing wrong with spam. But now, 'spam has gone too far and the quality of spam has become absolutely disgusting,' he said, adding, 'I take some of the blame for allowing it to go as far as it went.'"

Just days later, a Pennsylvania court found Sanford Wallace guilty of sending unsolicited junk faxes in violation of a 1991 federal law.

The Spam King went on to start up a number of Internet ventures, always claiming to be working against the evils of junk mail, and always somehow providing a substitute of equal power and charm, though of smaller profit margin.

The King's days on the throne may be over, but his influence shines still with the heat of a million bits of glitter. Sanford Wallace opened the door for new generations of spammers, whose clever technical innovations would bring far greater volumes of junk mail, in dazzling variety, to the peoples of the many nations of the earth. He showed them that it could be done, and that probably no one could stop them.

Up Theirs

"The surgical management of two patients presenting with incarcerated, apparently self-inserted foreign bodies is reported. The large volume of prior literature on this subject is reviewed, with tabulation of 182 previous cases by type and number of objects recovered and with a discussion of patients' age distribution, history, complications, and prognosis."

–from the medical report written by David B. Busch and James R. Starling

THE OFFICIAL CITATION

THE IG NOBEL LITERATURE PRIZE WAS AWARDED TO

David B. Busch and James R. Starling, of Madison, Wisconsin, for their deeply penetrating research report, "Rectal Foreign Bodies: Case Reports and a Comprehensive Review of the World's Literature." The citations include reports of, among other items: seven light bulbs; a knife sharpener; two flashlights; a wire spring; a snuff box; an oil can with potato stopper; eleven different forms of fruits, vegetables and other foodstuffs; a jeweler's saw; a frozen pig's tail; a tin cup; a beer glass; and one patient's remarkable ensemble collection consisting of spectacles, a suitcase key, a tobacco pouch, and a magazine.

Their study was published in *Surgery*, vol. 100, no. 3, September, 1986, pp. 512–19.

Most physicians have treated at least a handful of stunning cases – patients with ailments so striking, so wondrous as to merit a place in medical history. A surprising number of these cases involve people who have objects lodged in their rectums.

Dr James R. Starling treated several such cases, which so inspired him that, together with a colleague, he dug deep into the medical literature in search of more. The two doctors discovered much that had been hidden from the public. As a service to their profession, they wrote a comprehensive description – the world's first – of everything they'd found.

Rectal foreign bodies: Case reports and a comprehensive review of the world's literature

David B. Busch, Ph.D., M.D., *and* James R. Starling, M.D., *Madison, Wis.*

The surgical management of two patients presenting with incarcerated, apparently self-inserted foreign bodies is reported. The large volume of prior literature on this subject is reviewed, with tabulation of 182 previous cases by type and number of objects recovered and with a discussion of patients' age distribution, history, complications, and prognosis. Management problems addressed include history, differential diagnosis of reported pruitis ani, and handling of suspected assault. The variety of surgical techniques used to remove rectal foreign bodies transanally or after celiotomy is discussed. Vaginal foreign bodies and large bowel injuries due to fist fornication, colorectal instrumentation, pneumatic rupture, foreign body ingestion, impalement, and abdominal trauma are also discussed.

From the Departments of Pathology and Surgery, University of Wisconsin Hospital and Clinics, Madison, Wis.

INJURIES TO THE colon, rectum, and anus are important causes of morbidity and death. Among the sources of such injuries is the introduction of foreign bodies into the rectum in the absence of medical advice or approval. We report two cases of nontherapeutic introduction of rectal foreign bodies and review the substantial

removed by manual or endoscopic means. The patient consented to extraction of the dildo under general anesthetic. Biopsy specimens of the hemorrhagic rectal mucosa were performed and were negative on Ziehl-Neelson stains for mycobacterial or cryptosporidium infection. The patient was discharged without complications the following day.

Busch and Starling's Prize-winning report.

The case which first and best inspired Dr James R. Starling of Madison, Wisconsin, was quite unexpected:

"A 39-year-old married white male lawyer presented with a self-inserted perfume bottle in his rectum that he was unable to remove using various objects, including a back scratcher."

After a second instance of buried treasure presented itself at his office, Dr Starling enlisted his friend Dr David B. Busch in what was to become a classic of scholarly and medical inquiry. In their visits to medical libraries, Drs Busch and Starling went where no physicians had systematically gone before, poring through what they describe as "the substantial literature on this subject, including approximately 700 identified objects recovered from approximately 200 patients."

They took note of a 1937 *Kentucky Medical Journal* report that described the "insertion of a light bulb into a 52-year-old grandfather by several inebriated 'friends.'"

They puzzled over a 1959 write-up in the *South African*

Medical Journal detailing the "insertion of spectacles, a suitcase key, a tobacco pouch, and a magazine into a 38-year-old man by a 'friend.'"

Their attention was caught by a "case of suspected misreporting of an assault" described in a 1934 *New York State Journal of Medicine* article: "a 54-year-old married man admitted to self-insertion of two apples, having previously complained of assault by several men involving forced insertion of a vegetable (one cucumber and one parsnip)."

Drs Busch and Starling explain that in some cases, patients with this type of complaint will misreport certain aspects of the data. "This appears to be a means of coping with the embarrassment of the interview," they write. "Such patients should be treated with the utmost concern and tact, keeping in mind the great embarrassment they

Doctors have removed a variety of objects from people's rectums.
Photo: A.S. Kaswell/*Annals of Improbable Research*.

feel." A 1928 *American Journal of Surgery* article described a case of this sort: a "patient who initially admitted to self-insertion of a lemon and a cold cream jar and during his convalescence further stated that a drug clerk had advised him to use lemon juice and cold cream for relief of hemorrhoids, which were not found on examination." A 1935 report in the same journal concerned a patient who "presented with a broken broom handle, stating that he was using the object to massage his own prostate, a service allegedly rendered twice a week by his physician when the patient had more money." In 1932, at the height of the Great Depression, *The Illinois Medical Journal* described a patient who "reported self-insertion of two drinking glasses for relief of itching."

In addition to being a work of great literary and historical merit, Dr Busch and Dr Starling's "Rectal Foreign Bodies: Case Reports and a Comprehensive Review of the World's Literature" provides workaday technical information for their professional peers. The paper describes some of the means whereby objects, once found, can be successfully removed from their temporary abode.

"Light bulbs have been safely removed by padding the glass bulb with fine mesh gauze or cheesecloth followed by deliberate shattering of the object. Other ingenious mechanisms to remove light bulbs include a threaded broom handle and two large kitchen spoons.

"In one instance, a drinking glass was removed by packing the rectum with plaster of Paris to include an anchoring rope after the plaster of Paris had set.

"One particularly celebrated 16th century case involved a woman with a pig's tail inserted high in her rectum with the bristles directed caudad [i.e., pointing towards the

Here is the complete list compiled in 1986 by David B. Busch and James R. Starling of objects found in patients' rectums.

OBJECT	NO. RECOVERED
GLASS OR CERAMIC	
Bottle or jar	31
Bottle with attached rope	1
Glass or cup	12
Light bulb	7
Tube	6
FOOD	
Apple	1
Banana	2
Carrot	4
Cucumber	3
Onion	2
Parsnip	1
Plantain (with condom)	1
Potato	1
Salami	1
Turnip	1
Zucchini	2
WOODEN	
Ax handle	1
Stick or broom handle	10
Miscellaneous or unspecified	3
SEXUAL DEVICE	
Vibrator	23*
Dildo	15
KITCHEN DEVICE	
Dull knife	1
Ice pick	1
Knife sharpener	1
Mortar pestle	2
Spatula (plastic)	1
Spoon	1
Tin cup	1
MISCELLANEOUS TOOLS	
Candle	1
Flashlight	2
Iron rod	1
Pen	2
Rubber tube	1
Screwdriver	1
Toothbrush	1
Wire spring	1

OBJECT	NO. RECOVERED
INFLATED DEVICE	
Balloon	1
Balloon attached to cylinder	1
Condom	1
BALL	
Baseball	2
Tennis ball	1
MISCELLANEOUS CONTAINERS	
Baby powder can	1
Candle box	1
Snuff box	1
MISCELLANEOUS	
Bottle cap	1
Cattle horn	3
Frozen pig's tail	1
"Kangaroo tumor"	1
Plastic rod	1
Stone	2
Toothbrush holder	1
Toothbrush package	1
Whip handle	2 *
COLLECTIONS (one case of each)	
2 Glass tubes	
72½ jeweler's saw	
Oil can with potato stopper	
Piece of wood, peanut	
Umbrella handle and enema tubing	
2 Glasses	
Phosphorus match ends (homicide)	
402 Stones	
Toolbox **	
2 Bars soap	
Beer glass and preserving pot	
Lemon and cold cream jar	
2 Apples	
Spectacles, suitcase key, tobacco pouch, and magazine	

* number may be larger (text unclear)

**inside a convict; contained saws and other items usable in escape attempts.

rear]. In this case a hollow reed was cleverly inserted over the tail, which allowed easy removal of both objects together."

For casting light on many things which had once been kept in dark, deep places, David B. Busch and James R. Starling were awarded the 1995 Ig Nobel Prize in the field of Literature.

The winners could not travel to the Ig Nobel Prize Ceremony, but Dr Starling sent a videotaped acceptance speech. In it, wearing full medical scrubs and speaking in a world-weary monotone, he said:

"I am very appreciative to have been chosen as this year's recipient of the award. I have to, of course, thank the people I work with – Dr Busch, who encouraged me to give him my series of patients and was very helpful in the literature search. The reason that I'm wearing what I'm wearing is to remind any of you who wish to enter this sometimes dangerous field to make sure you're properly dressed, because the consequences can sometimes be somewhat devastating and a great surprise. If you're planning to do this kind of work, dress appropriately, have a sense of humor, and good luck."

Dr Busch and Dr Starling's report chronicled the panoply of objects that had been discovered *in situ* prior to 1986, its date of publication. Those who read the report realized that this was only a beginning.

In subsequent years, as consumer confidence soared, so, too, did the purchasing of goods that would find their way into people's rectums. Here is an arbitrary list of highlights from later years.

- 1987: The *American Journal of Forensic Medicine and Pathology* published a report entitled "Rectal Impaction Following Enema with Concrete Mix."

• 1991: The Japanese medical journal *Nippon Hoigaku Zasshi* detailed an unhappy case of "Homicide by Rectal Insertion of a Walking Stick."
• 1994: The *American Journal of Gastroenterology* reported a case of "Toothpick in Ano."
• 1996: The *Indian Journal of Gastroenterology* published a case report entitled "Whisky Bottle in the Rectum." The next year, that same journal published two reports of interest to the specialist. The first was called "Screwing a Carrot Out of the Rectum." The other was titled "And Now, a Needle in the Rectum."
• 1999: The *Journal of Emergency Medicine* reported the case of an oven mitt discovered inside the rectum of a 20-year-old man.
• 2001: The medical journal *Rozhledy v Chirurgii* reported that a porcelain cup was found in a gentleman in the Czech Republic; and the *British Dental Journal*, in a report titled "Don't Forget Your Toothbrush!" chronicled the case of a patient who did.

Appendices

- Year-by-Year List of Winners
- The Web Site
- About the *Annals of Improbable Research*

Year-by-Year List of Winners

1991

ECONOMICS – 1991

Michael Milken, titan of Wall Street and father of the junk bond, to whom the world is indebted.

PEACE – 1991

Edward Teller, father of the hydrogen bomb and first champion of the Star Wars weapons system, for his lifelong efforts to change the meaning of peace as we know it.

BIOLOGY – 1991

Robert Klark Graham, selector of seeds and prophet of propagation, for his pioneering development of the Repository for Germinal Choice, a sperm bank that accepts donations only from Nobellians and Olympians.

CHEMISTRY – 1991

Jacques Benveniste, prolific proseletizer and dedicated correspondent of *Nature*, for his persistent discovery that water, H_2O, is an intelligent liquid, and for demonstrating to his satisfaction that water is able to remember events long after all trace of those events has vanished.

MEDICINE – 1991

Alan Kligerman, deviser of digestive deliverance, vanquisher of vapor, and inventor of Beano, for his pioneering work with anti-gas liquids that prevent bloat, gassiness, discomfort and embarrassment.

EDUCATION – 1991

J. Danforth Quayle, consumer of time and occupier of space, for demonstrating, better than anyone else, the need for science education.

LITERATURE – 1991

Erich Von Däniken, visionary raconteur and author of *Chariots of the Gods?*, for explaining how human civilization was influenced by ancient astronauts from outer space.

1992

ECONOMICS – 1992

The investors of Lloyd's of London, heirs to 300 years of dull, prudent management, for their bold attempt to insure disaster by refusing to pay for their company's losses.

PEACE – 1992

Daryl Gates, former Police Chief of the City of Los Angeles, for his uniquely compelling methods of bringing people together.

BIOLOGY – 1992

Dr Cecil Jacobson, relentlessly generous sperm donor, and prolific patriarch of sperm banking, for devising a simple, single-handed method of quality control.

ARCHEOLOGY – 1992

Les Eclaireurs de France, the Protestant youth group whose name means "those who show the way," fresh-scrubbed removers of graffiti, for erasing the ancient paintings from the walls of the Mayrières Cave near the French village of Bruniquel.

PHYSICS – 1992

David Chorley and Doug Bower, lions of low-energy physics, for their circular contributions to field theory based on the geometrical destruction of English crops.

ART – 1992

Presented jointly to Jim Knowlton, modern Renaissance man, for his classic anatomy poster *Penises of the Animal Kingdom*, and to the US National Endowment for the Arts for encouraging Mr Knowlton to extend his work in the form of a pop-up book.

MEDICINE – 1992

F. Kanda, E. Yagi, M. Fukuda, K. Nakajima, T. Ohta and O. Nakata of the Shisedo Research Center in Yokohama, for their pioneering research study "Elucidation of Chemical Compounds Responsible for Foot Malodour," especially for their conclusion that people who think they have foot odor do, and those who don't, don't.

CHEMISTRY – 1992

Ivette Bassa, constructor of colorful colloids, for her role in the crowning achievement of 20th-century chemistry, the synthesis of bright blue Jell-O.

NUTRITION – 1992

The utilizers of Spam, courageous consumers of canned comestibles, for 54 years of undiscriminating digestion.

LITERATURE – 1992

Yuri Struchkov, unstoppable author from the Institute of Organoelemental Compounds in Moscow, for the 948 scientific papers he published between the years 1981 and 1990, averaging more than one every 3.9 days.

1993

ECONOMICS – 1993

Ravi Batra of Southern Methodist University, shrewd economist and best-selling author of *The Great Depression of 1990* ($17.95) and *Surviving the Great Depression of 1990* ($18.95), for selling enough copies of his books to single-handedly prevent worldwide economic collapse.

PEACE – 1993

The Pepsi-Cola Company of the Philippines, suppliers of sugary hopes and dreams, for sponsoring a contest to create a millionaire, and then announcing the wrong winning number, thereby inciting and uniting 800,000 riotously expectant winners, and bringing many warring factions together for the first time in their nation's history.

MEDICINE – 1993

James F. Nolan, Thomas J. Stillwell, and John P. Sands, Jr, medical men of mercy, for their painstaking research report, "Acute Management of the Zipper-Entrapped Penis."

PHYSICS – 1993

Louis Kervran of France, ardent admirer of alchemy, for his conclusion that the calcium in chickens' eggshells is created by a process of cold fusion.

CONSUMER ENGINEERING – 1993

Ron Popeil, incessant inventor and perpetual pitchman of late-night television, for redefining the industrial revolution with such devices as the Veg-O-Matic, the Pocket Fisherman, Mr Microphone, and the Inside-the-Shell Egg Scrambler.

VISIONARY TECHNOLOGY – 1993

Presented jointly to Jay Schiffman of Farmington Hills, Michigan, crack inventor of AutoVision, an image projection device that makes it possible to drive a car and watch television at the same time, and to the Michigan state legislature, for making it legal to do so.

MATHEMATICS – 1993

Robert Faid of Greenville, South Carolina, farsighted and faithful seer of statistics, for calculating the exact odds (710,609,175,188,282,000 to 1) that Mikhail Gorbachev is the Antichrist.

CHEMISTRY – 1993

James Campbell and Gaines Campbell of Lookout Mountain, Tennessee, dedicated deliverers of fragrance, for inventing scent strips, the odious method by which perfume is applied to magazine pages.

BIOLOGY – 1993

Paul Williams Jr, of the Oregon State Health Division, and

Kenneth W. Newell of the Liverpool School of Tropical Medicine, bold biological detectives, for their pioneering study, "Salmonella Excretion in Joy-Riding Pigs."

PSYCHOLOGY – 1993

John Mack of Harvard Medical School and David Jacobs of Temple University, mental visionaries, for their leaping conclusion that people who believe they were kidnapped by aliens from outer space, probably were – and especially for their conclusion that "the focus of the abduction is the production of children."

LITERATURE – 1993

E. Topol, R. Califf, F. Van de Werf, P. W. Armstrong, and their 972 co-authors, for publishing a medical research paper which has one hundred times as many authors as pages.

1994

MEDICINE – 1994

This prize is awarded in two parts. First, to Patient X, formerly of the US Marine Corps, valiant victim of a venomous bite from his pet rattlesnake, for his determined use of electroshock therapy: at his own insistence, automobile spark-plug wires were attached to his lip, and the car engine revved to 3,000 rpm for five minutes. Second, to Dr Richard C. Dart of the Rocky Mountain Poison Center and Dr Richard A. Gustafson of the University of Arizona Health Sciences Center, for their well-grounded medical report: "Failure of Electric Shock Treatment for Rattlesnake Envenomation."

PSYCHOLOGY – 1994

Lee Kuan Yew, former Prime Minister of Singapore, practitioner of the psychology of negative reinforcement, for his 30-year study of the effects of punishing three million citizens of Singapore whenever they spat, chewed gum, or fed pigeons.

ECONOMICS – 1994

Juan Pablo Davila of Chile, tireless trader of financial futures and former employee of the state-owned Codelco Company, for instructing his computer to "buy" when he meant "sell," and subsequently attempting to recoup his losses by making increasingly unprofitable trades that ultimately lost 0.5% of Chile's gross national product. Davila's relentless achievement inspired his countrymen to coin a new verb: " davilar," meaning, "to botch things up royally."

PEACE – 1994

John Hagelin of Maharishi University and the Institute of Science, Technology and Public Policy, promulgator of peaceful thoughts, for his experimental conclusion that 4,000 trained meditators caused an 18% decrease in violent crime in Washington, DC.

ENTOMOLOGY – 1994

Robert A. Lopez of Westport, New York, valiant veterinarian and friend of all creatures great and small, for his series of experiments in obtaining ear mites from cats, inserting them into his own ear, and carefully observing and analyzing the results.

PHYSICS – 1994

The Japan Meteorological Agency, for its seven-year study of whether earthquakes are caused by catfish wiggling their tails.

MATHEMATICS – 1994

The Southern Baptist Church of Alabama, mathematical measurers of morality, for their county-by-county estimate of how many Alabama citizens will go to Hell if they don't repent.

BIOLOGY – 1994

W. Brian Sweeney, Brian Krafte-Jacobs, Jeffrey W. Britton, and Wayne Hansen, for their breakthrough study, "The Constipated Serviceman: Prevalence Among Deployed US Troops," and especially for their numerical analysis of bowel movement frequency.

CHEMISTRY – 1994

Texas State Senator Bob Glasgow, wise writer of logical legislation, for sponsoring the 1989 drug control law which makes it illegal to purchase beakers, flasks, test tubes, or other laboratory glassware without a permit.

LITERATURE – 1994

L. Ron Hubbard, ardent author of science fiction and founding father of Scientology, for his crackling Good Book, *Dianetics*, which is highly profitable to mankind or to a portion thereof.

1995

PUBLIC HEALTH – 1995

Martha Kold Bakkevig of Sintef Unimed in Trondheim, Norway, and Ruth Nielson of the Technical University of Denmark, for their exhaustive study, "Impact of Wet Underwear on Thermoregulatory Responses and Thermal Comfort in the Cold."

DENTISTRY – 1995

Robert H. Beaumont, of Shoreview, Minnesota, for his incisive study "Patient Preference for Waxed or Unwaxed Dental Floss."

MEDICINE - 1995

Marcia E. Buebel, David S. Shannahoff-Khalsa, and Michael R. Boyle, for their invigorating study entitled "The Effects of Unilateral Forced Nostril Breathing on Cognition."

ECONOMICS – 1995

Awarded jointly to Nick Leeson and his superiors at Barings Bank and to Robert Citron of Orange County, California, for using the calculus of derivatives to demonstrate that every financial institution has its limits.

PEACE – 1995

The Taiwan National Parliament, for demonstrating that politicians gain more by punching, kicking and gouging each other than by waging war against other nations.

PSYCHOLOGY – 1995

Shigeru Watanabe, Junko Sakamoto, and Masumi Wakita, of Keio University, for their success in training pigeons to discriminate between the paintings of Picasso and those of Monet.

CHEMISTRY – 1995

Bijan Pakzad of Beverly Hills, for creating DNA Cologne and DNA Perfume, neither of which contain deoxyribonucleic acid, and both of which come in a triple helix bottle.

PHYSICS – 1995

D.M.R. Georget, R. Parker, and A.C. Smith, of the Institute of Food Research, Norwich, England, for their rigorous analysis of soggy breakfast cereal, published in the report entitled "A Study of the Effects of Water Content on the Compaction Behaviour of Breakfast Cereal Flakes."

NUTRITION – 1995

John Martinez of J. Martinez & Company in Atlanta, for luak coffee, the world's most expensive coffee, which is made from coffee beans ingested and excreted by the luak (a.k.a., the palm civet), a bobcat-like animal native to Indonesia.

LITERATURE – 1995

David B. Busch and James R. Starling, of Madison, Wisconsin, for their deeply penetrating research report, "Rectal foreign bodies: Case Reports and a Comprehensive Review of the World's Literature." The citations include reports of, among other items: seven light bulbs; a knife sharpener; two flashlights; a wire spring; a snuff box; an oil can with potato stopper; 11 different forms of fruits, vegetables and other foodstuffs; a jeweler's saw; a frozen pig's tail; a tin cup; a beer glass; and one patient's remarkable ensemble collection consisting of spectacles, a suitcase key, a tobacco pouch, and a magazine.

1996

PUBLIC HEALTH – 1996

Ellen Kleist of Nuuk, Greenland, and Harald Moi of Oslo, Norway, for their cautionary medical report "Transmission of Gonorrhea Through an Inflatable Doll."

MEDICINE – 1996

James Johnston of R.J. Reynolds, Joseph Taddeo of US Tobacco, Andrew Tisch of Lorillard, William Campbell of Philip Morris, Edward A. Horrigan of Liggett Group, Donald S. Johnston of American Tobacco Company, and the late Thomas E. Sandefur, Jr, chairman of Brown and Williamson Tobacco Co. for their unshakable discovery, as testified to the US Congress, that nicotine is not addictive.

ECONOMICS – 1996

Dr Robert J. Genco of the University of Buffalo for his discovery that "financial strain is a risk indicator for destructive periodontal disease."

PEACE – 1996

Jacques Chirac, President of France, for commemorating the 50th anniversary of Hiroshima with atomic bomb tests in the Pacific.

BIODIVERSITY – 1996

Chonosuke Okamura of the Okamura Fossil Laboratory in Nagoya, Japan, for discovering the fossils of dinosaurs, horses, dragons, princesses, and more than 1,000 other extinct "mini-species," each of which is less than 1⁄100 of an inch in length.

PHYSICS – 1996

Robert Matthews of Aston University, England, for his studies of Murphy's Law, and especially for demonstrating that toast often falls on the buttered side.

ART – 1996

Don Featherstone of Fitchburg, Massachusetts, for his

ornamentally evolutionary invention, the plastic pink flamingo.

CHEMISTRY – 1996

George Goble of Purdue University, for his blistering world record time for igniting a barbecue grill – three seconds, using charcoal and liquid oxygen.

BIOLOGY – 1996

Anders Baerheim and Hogne Sandvik of the University of Bergen, Norway, for their tasty and tasteful report, "Effect of Ale, Garlic, and Soured Cream on the Appetite of Leeches."

LITERATURE – 1996

The editors of the journal *Social Text*, for eagerly publishing research that they could not understand, that the author said was meaningless, and which claimed that reality does not exist.

1997

PEACE – 1997

Harold Hillman of the University of Surrey, England, for his lovingly rendered and ultimately peaceful report "The Possible Pain Experienced During Execution by Different Methods."

MEDICINE – 1997

Carl J. Charnetski and Francis X. Brennan, Jr. of Wilkes University, and James F. Harrison of Muzak Ltd in Seattle, Washington, for their discovery that listening to elevator Muzak stimulates immunoglobulin A (IgA) production, and thus may help prevent the common cold.

BIOLOGY – 1997

T. Yagyu and his colleagues from the University Hospital of Zurich, Switzerland, from Kansai Medical University in Osaka, Japan, and from Neuroscience Technology Research in Prague, Czech Republic, for measuring people's brainwave patterns while they chewed different flavors of gum.

ECONOMICS – 1997

Akihiro Yokoi of Wiz Company in Chiba, Japan, and Aki Maita of Bandai Company in Tokyo, the father and mother of Tamagotchi, for diverting millions of person-hours of work into the husbandry of virtual pets.

ENTOMOLOGY – 1997

Mark Hostetler of the University of Florida, for his scholarly book, *That Gunk on Your Car*, which identifies the insect splats that appear on automobile windows.

ASTRONOMY – 1997

Richard Hoagland of New Jersey, for identifying artificial features on the moon and on Mars, including a human face on Mars and ten-mile-high buildings on the far side of the moon.

PHYSICS – 1997

John Bockris of Texas A&M University, for his wide-ranging achievements in cold fusion, in the transmutation of base elements into gold, and in the electrochemical incineration of domestic rubbish.

METEOROLOGY – 1997

Bernard Vonnegut of the State University of Albany, for his revealing report, "Chicken Plucking as Measure of Tornado Wind Speed."

LITERATURE – 1997

Doron Witztum, Eliyahu Rips and Yoav Rosenberg of Israel, and Michael Drosnin of the United States, for their hairsplitting statistical discovery that the Bible contains a secret, hidden code.

COMMUNICATIONS – 1997

Sanford Wallace, president of Cyber Promotions of Philadelphia – neither rain nor sleet nor dark of night have stayed this self-appointed courier from delivering electronic junk mail to all the world.

1998

PEACE – 1998

Prime Minister Shri Atal Bihari Vajpayee of India and Prime Minister Nawaz Sharif of Pakistan, for their aggressively peaceful explosions of atomic bombs.

ECONOMICS – 1998

Richard Seed of Chicago for his efforts to stoke up the world economy by cloning himself and other human beings.

STATISTICS – 1998

Jerald Bain of Mt Sinai Hospital in Toronto and Kerry Siminoski of the University of Alberta for their carefully measured report, "The Relationship Among Height, Penile Length, and Foot Size."

BIOLOGY – 1998

Peter Fong of Gettysburg College, Gettysburg, Pennsylvania, for contributing to the happiness of clams by giving them Prozac.

CHEMISTRY – 1998

Jacques Benveniste of France, for his homeopathic discovery that not only does water have memory, but that the information can be transmitted over telephone lines and the Internet.

SAFETY ENGINEERING – 1998

Troy Hurtubise, of North Bay, Ontario, for developing, and personally testing a suit of armor that is impervious to grizzly bears.

MEDICINE – 1998

To Patient Y and to his doctors, Caroline Mills, Meirion Llewelyn, David Kelly, and Peter Holt, of Royal Gwent Hospital, in Newport, Wales, for the cautionary medical report, "A Man Who Pricked His Finger and Smelled Putrid for 5 Years."

SCIENCE EDUCATION – 1998

Dolores Krieger, Professor Emerita, New York University, for demonstrating the merits of Therapeutic Touch, a method by which nurses manipulate the energy fields of ailing patients by carefully avoiding physical contact with those patients.

PHYSICS – 1998

Deepak Chopra of the Chopra Center for Well Being, La Jolla, California, for his unique interpretation of quantum physics as it applies to life, liberty, and the pursuit of economic happiness. (Deepak Chopra's books *Quantum Healing*, *Ageless Body, Timeless Mind*, etc.)

LITERATURE – 1998

Dr Mara Sidoli of Washington, DC, for her illuminating report, "Farting as a Defence Against Unspeakable Dread."

1999

PEACE – 1999

Charl Fourie and Michelle Wong of Johannesburg, South Africa, for inventing an automobile burglar alarm consisting of a detection circuit and a flame-thrower.

CHEMISTRY – 1999

Takeshi Makino, president of The Safety Detective Agency in Osaka, Japan, for his involvement with S-Check, an infidelity detection spray that wives can apply to their husbands' underwear.

MANAGED HEALTH CARE – 1999

The late George and Charlotte Blonsky of New York City and San Jose, California, for inventing a device (US Patent #3,216,423) to aid women in giving birth – the woman is strapped onto a circular table, and the table is then rotated at high speed.

ENVIRONMENTAL PROTECTION – 1999

Hyuk-ho Kwon of Kolon Company of Seoul, Korea, for inventing the self-perfuming business suit.

BIOLOGY – 1999

Dr Paul Bosland, director of The Chile Pepper Institute, New Mexico State University, Las Cruces, New Mexico, for breeding a spiceless jalapeño chile pepper.

LITERATURE – 1999

The British Standards Institution for its six-page specification (BS 6008) of the proper way to make a cup of tea.

SOCIOLOGY – 1999

Steve Penfold, of York University in Toronto, for doing his PhD thesis on the sociology of Canadian donut shops.

PHYSICS – 1999

Dr Len Fisher of Bath, England, and Sydney, Australia, for calculating the optimal way to dunk a biscuit
... and ...
Professor Jean-Marc Vanden-Broeck of the University of East Anglia, England, and Belgium, for calculating how to make a teapot spout that does not drip.

MEDICINE – 1999

Dr Arvid Vatle of Stord, Norway, for carefully collecting, classifying, and contemplating which kinds of containers his patients chose when submitting urine samples.

SCIENCE EDUCATION – 1999

The Kansas State Board of Education and the Colorado State Board of Education, for mandating that children should not believe in Darwin's theory of evolution any more than they believe in Newton's theory of gravitation, Faraday's and Maxwell's theory of electromagnetism, or Pasteur's theory that germs cause disease.

2000

PSYCHOLOGY – 2000

David Dunning of Cornell University and Justin Kreuger of the University of Illinois, for their modest report, "Unskilled and

Unaware of It: How Difficulties in Recognizing One's Own Incompetence Lead to Inflated Self-Assessments."

PEACE – 2000

The British Royal Navy, for ordering its sailors to stop using live cannon shells, and instead just to shout "Bang!"

CHEMISTRY – 2000

Donatella Marazziti, Alessandra Rossi, and Giovanni B. Cassano of the University of Pisa, and Hagop S. Akiskal of the University of California (San Diego), for their discovery that, biochemically, romantic love may be indistinguishable from having severe obsessive-compulsive disorder.

ECONOMICS – 2000

The Reverend Sun Myung Moon, for bringing efficiency and steady growth to the mass-marriage industry, with, according to his reports, a 36-couple wedding in 1960, a 430-couple wedding in 1968, an 1,800-couple wedding in 1975, a 6,000-couple wedding in 1982, a 30,000-couple wedding in 1992, a 360,000-couple wedding in 1995, and a 36,000,000-couple wedding in 1997.

MEDICINE – 2000

Willibrord Weijmar Schultz, Pek van Andel, and Eduard Mooyaart of Groningen, The Netherlands, and Ida Sabelis of Amsterdam, for their illuminating report, "Magnetic Resonance Imaging of Male and Female Genitals During Coitus and Female Sexual Arousal."

PUBLIC HEALTH – 2000

Jonathan Wyatt, Gordon McNaughton, and William Tullet of Glasgow, for their alarming report, "The Collapse of Toilets in Glasgow."

PHYSICS – 2000

Andre Geim of the University of Nijmegen, The Netherlands, and Sir Michael Berry of Bristol University, England, for using magnets to levitate a frog.

COMPUTER SCIENCE – 2000

Chris Niswander of Tucson, Arizona, for inventing PawSense, software that detects when a cat is walking across your computer keyboard.

BIOLOGY – 2000

Richard Wassersug of Dalhousie University, for his first-hand report, "On the Comparative Palatability of Some Dry-Season Tadpoles from Costa Rica."

LITERATURE – 2000

Jasmuheen (formerly known as Ellen Greve) of Australia, first lady of Breatharianism, for her book *Living on Light*, which explains that although some people do eat food, they don't ever really need to.

2001

PUBLIC HEALTH – 2001

Chittaranjan Andrade and B.S. Srihari of the National Institute of Mental Health and Neurosciences, Bangalore, India, for their probing medical discovery that nose picking is a common activity among adolescents.

PSYCHOLOGY – 2001

Lawrence W. Sherman of Miami University, Ohio, for his influential research report "An Ecological Study of Glee in Small Groups of Preschool Children."

ECONOMICS – 2001

Joel Slemrod, of the University of Michigan Business School, and Wojciech Kopczuk, of the University of British Columbia, for their conclusion that people find a way to postpone their deaths if that would qualify them for a lower rate on the inheritance tax.

PEACE – 2001

Viliumas Malinauskas of Grutas, Lithuania, for creating the amusement park known as "Stalin World."

MEDICINE – 2001

Peter Barss of McGill University, for his impactful medical report "Injuries Due to Falling Coconuts."

PHYSICS – 2001

David Schmidt of the University of Massachusetts for his partial solution to the question of why shower curtains billow inwards.

TECHNOLOGY – 2001

Awarded jointly to John Keogh of Hawthorn, Victoria, Australia, for patenting the wheel in the year 2001, and to the Australian Patent Office for granting him Innovation Patent #2001100012.

ASTROPHYSICS – 2001

Dr Jack and Rexella Van Impe of Jack Van Impe Ministries, Rochester Hills, Michigan, for their discovery that black holes fulfill all the technical requirements to be the location of Hell.

BIOLOGY – 2001

Buck Weimer of Pueblo, Colorado, for inventing Under-Ease, air-tight underwear with a replaceable charcoal filter that removes bad-smelling gases before they escape.

LITERATURE – 2001

John Richards of Boston, England, founder of The Apostrophe Protection Society, for his efforts to protect, promote, and defend the differences between plural and possessive.

2002

BIOLOGY – 2002

Norma E. Bubier, Charles G.M. Paxton, Phil Bowers, and D. Charles Deeming of the United Kingdom, for their report "Courtship Behaviour of Ostriches Towards Humans Under Farming Conditions in Britain."

PHYSICS – 2002

Arnd Leike of the University of Munich, for demonstrating that beer froth obeys the mathematical Law of Exponential Decay.

INTERDISCIPLINARY RESEARCH – 2002

Karl Kruszelnicki of The University of Sydney, for performing a comprehensive survey of human belly button lint – who gets it, when, what color, and how much.

CHEMISTRY – 2002

Theodore Gray of Wolfram Research, in Champaign, Illinois, for gathering many elements of the periodic table, and assembling them into the form of a four-legged periodic table table.

MATHEMATICS – 2002

K.P. Sreekumar and the late G. Nirmalan of Kerala Agricultural University, India, for their analytical report "Estimation of the Total Surface Area in Indian Elephants."

LITERATURE – 2002

Vicki L. Silvers of the University of Nevada-Reno and David S. Kreiner of Central Missouri State University, for their colorful report "The Effects of Pre-Existing Inappropriate Highlighting on Reading Comprehension."

PEACE – 2002

Keita Sato, President of Takara Co., Dr Matsumi Suzuki, President of Japan Acoustic Lab, and Dr Norio Kogure, Executive Director, Kogure Veterinary Hospital, for promoting peace and harmony between the species by inventing Bow-Lingual, a computer-based automatic dog-to-human language translation device.

HYGIENE – 2002

Eduardo Segura, of Lavakan de Aste, in Tarragona, Spain, for inventing a washing machine for cats and dogs.

ECONOMICS – 2002

The executives, corporate directors, and auditors of Enron, Lernout & Hauspie, Adelphia, Bank of Commerce and Credit International, Cendant, CMS Energy, Duke Energy, Dynegy, Gazprom, Global Crossing, HIH Insurance, Informix, Kmart, Maxwell Communications, McKessonHBOC, Merrill Lynch, Merck, Peregrine Systems, Qwest Communications, Reliant Resources, Rent-Way, Rite Aid, Sunbeam, Tyco, Waste Management, WorldCom, Xerox, and Arthur Andersen, for adapting the mathematical concept of imaginary numbers for use in the business world.

MEDICINE – 2002

Chris McManus of University College London, for his excruciatingly balanced report, "Scrotal Asymmetry in Man and in Ancient Sculpture."

The Web Site

The Ig Nobel Prizes home page is at www.improbable.com

It is part of the *Annals of Improbable Research* web site.

There you will find a complete list of the winners and, in many cases, links to their home pages, their original research, and press clippings about them. You will also find video from some of the ceremonies, and links to the recordings of the annual Ig Nobel broadcast on National Public Radio's *Talk of the Nation/Science Friday with Ira Flatow* program. We also put up occasional news of the continuing adventures of past Ig Nobel Prize winners.

FREE NEWSLETTER: To keep informed of upcoming Ig Nobel ceremonies and related events, add yourself to the distribution list for the free monthly newsletter *mini-AIR*. You can do that at the web site or, alternatively:

Send a brief e-mail message to this address:

LISTPROC@AIR.HARVARD.EDU

The body of your message should contain *only* the words "SUBSCRIBE MINI-AIR" followed by your name. Here are two examples:

SUBSCRIBE MINI-AIR Irene Curie Joliot
SUBSCRIBE MINI-AIR Nicholai Lobachevsky

About the *Annals of Improbable Research*

The *Annals of Improbable Research* (AIR) is a humor magazine about science, medicine, and technology. It may be the only science journal read not just by scientists and doctors, but also by their families and friends.

AIR is known for:

- funny, genuine research, culled from somber science and medical journals;
- deadpan satire; and
- the Ig Nobel Prizes.

About a third of what we publish in *AIR* is genuine research, about a third is concocted, and about a third of our readers cannot tell the difference. (In the magazine we always indicate which items come from official sources – and we even give you the info to go look and see for yourself.)

Every year, we devote one issue to a full report on the Ig Nobel Prize Ceremony, with all the juicy details, lots of photos, and all the words for that year's Ig Nobel mini-opera. You can subscribe to the magazine (6 issues per year) by going to the web site (www.improbable.com), or by mail or telephone:

see overleaf for subscription form.

PLEASE ❑ START OR ❑ RENEW A SUBSCRIPTION FOR ME

❑ 1 YEAR (6 ISSUES) ❑ 2 YEARS (12 ISSUES)

❑ START OR ❑ RENEW A GIFT SUBSCRIPTION ❑ 1 YEAR (6 ISSUES)
❑ 2 YEARS (12 ISSUES)

MY NAME, ADDRESS AND ALL THAT

NAME ....

ADDRESS ....

....

....

....

....

....

....

PHONE ....

FAX ....

E-MAIL ....

I AM GIVING A SUBSCRIPTION TO

NAME ....

ADDRESS ....

....

....

....

....

....

....

PHONE ....

FAX ....

E-MAIL ....

❑ SEND RENEWAL NOTICE TO MY BENEFICIARY

❑ SEND RENEWAL NOTICE TO ME

RATES (IN US DOLLARS)

USA	$29	$53
CANADA/MEX	$33	$57
OVERSEAS	$45	$82

TOTAL PAYMENT ENCLOSED ....

PAYMENT METHOD

❑ CHECK (DRAWN ON US BANK) OR INTERNATIONAL MONEY ORDER

❑ MASTERCARD ❑ VISA

❑ DISCOVER

CARD NUMBER ....

EXP. DATE ....

Send payment to:
Annals of Improbable Research
PO Box 380853
Cambridge
MA 02238
USA

Telephone: (617) 491-4437
Fax: (617) 661-0927
E-mail: air@improbable.com

www.improbable.com